City in the Cloud – Data on the Ground
The Architecture of Data

City in the Cloud   Data on the Ground

The Architecture of Data

ARCHITANGLE

Prologue

6   *The Architecture of Data*
    Cara Hähl-Pfeifer, Damjan Kokalevski, and Andres Lepik

Elemental

18 Marina Otero Verzier in Conversation with Cara Hähl-Pfeifer, Damjan Kokalevski, Andres Lepik, and Māra Starka

24 *Submarine Cables: Colonial Roots of Global Communication*
    Photo Essay

42 *The Necropolitics of Big AI*
    Mél Hogan

50 *Lithium Mining in the Atacama Desert*
    Photo Essay by Catherine Hyland

68 *Exhuming Earth: Extraction and Resistance in the Age of the "Green Transition"*
    Godofredo Enes Pereira

Spatial

76 Marina Otero Verzier in Conversation with Cara Hähl-Pfeifer, Damjan Kokalevski, Andres Lepik, and Māra Starka

82 *Land and Power*
    Marija Marić

88 *Where the Cloud Becomes Reality: On the Iconography of the Data Center*
    Niklas Maak

96 *Leibniz Supercomputing Centre of the Bavarian Academy of Sciences and Humanities*
    Photo Essay by Giulia Bruno

114 *Undoing Optimization: Quiet Rewilding*
    Alison Powell

Temporal

122 Marina Otero Verzier in Conversation with Cara Hähl-Pfeifer, Damjan Kokalevski, Andres Lepik, and Māra Starka

128 *Constructing Cartographies: Mapping Territories through Time and Technology*
    Andra Pop-Jurj

142 *In a Manner of Speaking: From Bits to Bricks*
    Teresa Fankhänel and Max Hallinan

148 *Forgetting as a Feature, Not a Bug: The Intelligence of Loss in the Archive*
    Anna-Maria Meister and Rafael Uriarte

158 *Carrying Data*
    Photo Essay by Giulia Bruno

Epilogue

192 *Regaining Agency in a Technological Swamp*
    James Bridle in Conversation with Cara Hähl-Pfeifer and Damjan Kokalevski

Appendix

202 Biographies
206 Image Credits
207 Acknowledgments
208 Colophon

Cara Hähl-Pfeifer, Damjan Kokalevski, and Andres Lepik

The Architecture of Data

Data and technological development dominate our lives. Today, people enjoy almost unlimited access to information and communication in real time thanks to increasing computing power and ever-faster access to worldwide data storage connected via the internet. However, our ways of interacting with our digitally connected spaces have turned us into mere data sources. Most systems that produce, exchange, and process data remain hidden from view. They are only visible through the myriad devices and screens that populate our lives, like phones, laptops, smart home appliances, car interfaces, and countless displays in urban environments.

Hidden behind the scenes lie the systemic code—the software and algorithms that control data flows and processing—and the underlying material infrastructure, such as microchips, batteries, land and undersea cables, data centers, and server farms. Even more obscured is the human labor needed to keep these systems operational, along with the enormous energy demands required for seamless, real-time, and 24/7 availability and reliability of data. As the global data economy continues to grow, it increasingly depends on generating more energy and extracting more critical raw materials. Consequently, this puts enormous strain on the environment and endangers the livelihoods of local communities, making the quest for securing digital futures a political act.

While technological advancement promises a life of optimization, comfort, and new enlightenment, technology alone cannot be used as the foundation for societal and political progress. The latest technological fix—artificial intelligence (AI)—is often disproportionately set to benefit a few multinational corporations and Western countries, to the detriment and continuous exploitation of human labor and material resources from the Global South.[1] The enormous wealth created under these techno-capitalist conditions cannot be distributed fairly, and full automation will not simply "solve" this issue. It must start with a comprehensive building of infrastructure for the benefit and well-being of all.

According to the European Union (EU)'s Critical Raw Materials Act of 2024, the EU's ability to achieve its climate and digital objectives by 2030 is at high risk.

This is mainly due to strategic dependencies on critical raw materials relying heavily on imports, often from one country, such as China, which currently provides 100 percent of the EU's supply of heavy rare earth elements, or Chile, which provides 79 percent of the EU's supply of lithium. Faced with the expected exponential increase in demand for these materials, the EU has reevaluated its resource supply chain, focusing inward. The act emphasizes that securing and diversifying sources and mitigating strategic dependencies are crucial not only for realizing the EU's digital future, but also for ensuring European sovereignty.[2]

As the EU competes with the United States and China for securing data economies, it has to navigate a tightrope between adopting an aggressive strategy for extracting materials and building digital infrastructure while upholding European values of democratic governance, which often do not go hand in hand. This balancing act is particularly evident in the recent Artificial Intelligence Act, which seeks to position the EU as a leader in AI development while managing and controlling the risks associated with the development of AI. This act also aims to bring transparency and trust in the Union's administration of its AI futures,[3] a challenging task that is overshadowed by the looming threat of an "AI arms race."[4] In this polarized landscape, the dangers associated with the seemingly limitless expansion of digital infrastructure and its impacts on Earth are often deliberately left out of public debate and scrutiny.

Mapping as Agency

This publication serves as a guide to uncovering the material, spatial, and temporal aspects of data. It is structured as a three-part study that showcases academic and artistic research, challenging the harm inherent in data abstraction and aiming to create grounded knowledge about these complex infrastructures. By turning to interdisciplinary spatial and architectural research, this work reveals the links between architecture and agency.

Scan the QR code to see exhibition views and documentation of the public program.

The publication accompanies our exhibition *City in the Cloud – Data on the Ground*, on display at the Architecture Museum of the Technical University of Munich (Architekturmuseum der TUM), in the Pinakothek der Moderne, from October 16, 2025, to March 8, 2026. The exhibition is the result of two years of research conducted at TUM's Chair of Architecture History and Curatorial Practice, involving scholars and students from various fields.

The book is structured by a three-part conversation with main exhibition collaborator and researcher Marina Otero Verzier, linking it to the show's guiding themes. It expands upon these topics with contributions that range from academic essays and photo essays to archival stories and poetry by experts in architecture, the humanities, media studies, journalism, and art. The research presented aims to reveal the complex relationships between data, its structures, and the spaces it inhabits, fostering long-term engagement with these issues.

The first chapter outlines an *elemental* aspect of data, focusing on large-scale infrastructures and the extraction of raw materials essential for the continuous construction of digital infrastructure. The second chapter examines the *spatial* aspects of data, turning to spaces of data production and consumption, including a discussion on the architecture of data centers and a reflection on the mantra of the "smart city." The third chapter studies the *temporal* dimensions of data, discussing the limits of storage and the role of digital data in preserving and archiving architecture. This chapter also addresses the significance of AI and its relation to language and architecture histories, proposing agendas for interpreting datasets, their maintenance, and archiving.

The book and exhibition assert that architects, designers, and researchers already possess agency and emphasize the importance of curatorial research and mapping as powerful tools at their disposal. When used conscientiously, these tools become political instruments, reveal nuanced complexities, and address hidden issues such as dispossession and injustice. As Otero points out: "In that sense, the map becomes not just a tool of representation, but of inquiry—and potentially, of accountability."

Elemental

Digital systems, often rooted in (neo-)colonial oppressive practices, are Earth-bound and impact entire ecosystems and the lives of both humans and nonhumans on a planetary scale. In his essay "Exhuming Earth: Extraction and Resistance in the Age of the 'Green Transition,'" Godofredo Enes Pereira describes this as the extractive gaze, which unsees life itself to focus on the riches below. Following the production of one of the most sought-after minerals, lithium, and its implications in Chile and Portugal, Pereira suggests ways for both countries to learn and support one another in fostering environmental awareness, coupled with political will.

As these expansive strategies materialize, questions about how our data moves around the globe, where it is stored, and who has access to it become increasingly crucial. Data profoundly shapes our virtual and physical lives, affecting everything from media consumption to mobility and the design of our living and working spaces. As Big Tech and Big AI attempt to control every waking and dreaming facet of our lives, public trust in our digital systems is dwindling. With the transition to a global digital economy and "in the shadow of virtualization," the concept of the data body has emerged. Consisting of all digital traces linked to an individual, it has become similarly important to platform capitalism and rampant surveillance and policing as its physical counterpart.[5] Citizens are thereby stripped of their political agency and reduced to mere datasets for the profit of a handful of multinational corporations. Reducing citizens to data bodies enforces disempowerment, challenging almost every aspect of our society, including its democratic values, work, economy, and cultural production. In her essay "The *Necro*politics of Big AI," Mél Hogan warns that data development is disguised as a powerful political fiction. Following this argument, she links data infrastructures to the politics of technological fixes, discussing how these data systems are pushing our democracies to their limits.

Spatial

The architecture of data centers is scaled up from its smallest unit: the server rack. Consequently, the locations where our data is processed and stored remain abstract. Their environment and sensory landscapes quite literally repel humans. Giulia Bruno captures this sentiment in her artistic investigation of the Leibniz Supercomputing Centre in Munich (LRZ): "It is a hard-to-define smell, bitter and sharp, reminiscent of heated plastic and freshly welded metal, artificial, a combination of warm air and a technical world where matter bends to computation." The LRZ is a scientific institution that merges public infrastructure with societal responsibility around a central machine core that needs constant human maintenance. Compared to privately owned infrastructures like Equinix's colocation and hyperscale data centers—currently the largest data center provider globally—the LRZ exemplifies an unusually high level of design.

Typical data centers, often located on the outskirts of cities, on industrial or agricultural land, or in business parks, are designed to go unnoticed by the untrained eye. Their architecture resembles functional warehouses, with few openings to the outside. This is hardly surprising, given that a typical mid-sized data center employs no more than fifteen people at any given time. Examining the architecture of these data production facilities reveals their seemingly disconnected relationship with surrounding communities and the environment—they are often out of sight and out of mind. However, appearances can be misleading, as these infrastructures depend heavily on local supply systems, such as water, power, and labor. This also emphasizes how different ecologies of data require different and varying architectural responses. As Otero Verzier explains, these ecologies depend on our relationship with data and the value we place on it: "Cold and hot data, preserved data, and real-time data or data for training AI. If we understand those ecologies clearly," Otero Verzier argues, "they can lead to entirely different architectural approaches to data centers, ones that are more nuanced, responsive, and socially and ecologically meaningful."

This calls for more regulation and strategic planning of data centers, alongside a need for greater architectural innovation and interdisciplinary collaboration between data sciences and spatial sciences. Addressing the urban scale and the spaces within cities requires us to move beyond the current mantra of "smartness," which is based on continuous optimization and exploitation. In the essay "Undoing Optimization: Quiet Rewilding," Alison Powell advocates for a less tech-dependent environment. Rewilding is seen as a form of resistance in contrast to the notions of optimization and capitalist progress. This perspective means "observing things otherwise, making space for others and sitting with difference."

The various types of knowledge in our cities—both rational and embodied—are challenged by our dependency on digital data systems and the consequent loss of agency in the name of comfort, convenience, and user-friendliness. We delegate more computing and decision-making tasks to machines and, with the rise of AI, even integrate our very cognition into the cloud. We then encounter the black box, an opaque cloud—something you cannot experience, grasp, or see. As James Bridle concludes in "Regaining Agency in a Technological Swamp," architecture becomes agency because this very opaqueness contradicts the experiential view of space, that seeing something, like architecture, should be a bodily experience. "When it ceases to be bodily and physical, something is wrong."

Temporal

As data accumulates, a crucial question arises: What should we keep, and what can we let go of? How can we make data more manageable and ecologically sustainable? Systems designed to preserve our knowledge, such as archives, struggle to keep pace in an age of excessive backup and storage madness. These backup systems face threats of destruction, driven by both human and natural forces, including institutional policies, wars, migration, natural disasters, and decay, to name a few. Anna-Maria Meister and Rafael Uriarte argue that archives have "their very own flora and fauna." For example, if data has been forgotten, unprocessed, or not retrieved for a long period, does it even really exist? In their essay "Forgetting as a Feature, Not a Bug: The Intelligence of Loss in the Archive," Meister and Uriarte join forces to put forward an idea for an algorithmically driven and intelligent process of destroying data within the archive.

Realizing that digital and archival space is not infinite requires a more nuanced approach to *data curation*. There is a need for more awareness around data management, its archiving, and long-term strategies for determining what to keep and let go of, challenging the idea of meaningless accumulation. As Marina Otero Verzier states, "It's not necessarily the case that more information leads to more knowledge. … Especially now that we rely on infrastructures like artificial intelligence to provide us with information, the question becomes: How do we assess the value and accuracy of that information?"

The Architekturmuseum der TUM was formed around an extensive archive containing over 660,000 drawings, 200,000 photographs, and 1,500 models. While these items are still stored physically, they are increasingly being digitized and made available online. In 2020, the exhibition *The Architecture Machine*, curated by Teresa Fankhänel, prompted an investigation of the historical interactions between the development of the computer and architectural design. How have architects embraced and molded this new technology to their own ends? In light of the ever-advancing digitalization of architectural practices, we decided to step back and examine the broader historical context and material conditions on a global scale.

This project serves as a curatorial meditation on data and its infrastructure, introducing these topics into the architectural discourse through an interdisciplinary and media research perspective. By closely following the data infrastructure—from the cables deep in the oceans that encircle the globe to data centers, supercomputers, all the way to our homes, cities, and devices—a particular map emerges. This map helps demystify the structures that shape our virtual and physical lives. Bringing together diverse voices and media across disciplines, we hope the research, case studies, and stories presented will contribute to ongoing discussions about the significance of architecture, planning, and design in fostering democratic and eco-technological collective futures.

1 "Digitaler Kolonialismus Faktencheck," Brot für die Welt website, https://www.brot-fuer-die-welt.de/themen/digitalisierung/digitaler-kolonialismus-faktencheck/.

2 European Council, "Critical Raw Materials Act," https://www.consilium.europa.eu/en/infographics/critical-raw-materials/.

3 European Council, "AI Act | Shaping Europe's Digital Future," August 1, 2024, https://digital-strategy.ec.europa.eu/en/policies/regulatory-framework-ai.

4 Alvin Wang Graylin and Paul Triolo, "There Can Be No Winners in a US-China AI Arms Race," *MIT Technology Review*, January 21, 2025, https://www.technologyreview.com/2025/01/21/1110269/there-can-be-no-winners-in-a-us-china-ai-arms-race/.

5 Konrad Becker, "Globale Datenkörper," in *Die Politik der Infosphäre* (Wiesbaden: VS Verlag für Sozialwissenschaften, 2003), pp. 195–209, here p. 196.

Elemental

Marina Otero Verzier in Conversation with Cara Hähl-Pfeifer, Damjan Kokalevski,
Andres Lepik, and Māra Starka

Elemental

In the first part of the conversation, Marina Otero Verzier guides readers through the hidden
foundational elements of datascapes. She questions the seamlessness and efficiency that digital
infrastructure typically demands—aspects that often result in the concealment of these systems.
Arguing for the agency of architects, designers, and researchers, she introduces mapping as
a powerful and political tool to critically examine these systems while remaining sensitive to
local experiences and forms of resistance. The discussion then delves into extracting materials
needed to build digital infrastructure—focusing on the destructive practices of mining one of
the most sought-after minerals, lithium.

Cara Hähl-Pfeifer, Damjan Kokalevski, Andres Lepik, and Māra Starka: In your writings, you often reveal how closely our everyday lives are connected to spaces of extraction, an idea that directly emerges from using mapping as a research method. Mapping, however, is never neutral and is deeply rooted in colonial history. Given this, can it still serve as a tool to better understand the impact of data infrastructures, for example on our planet and communities? If so, what is the first link that we can make between extraction practices and our agency as architects when producing such maps?

Marina Otero Verzier: This form of cartography has historically been tied to ideas of sovereignty and control—tools for demarcating, claiming, and managing territory. And we're very used to working with these kinds of maps: highly detailed, yet fundamentally abstract. They show vast geographies and how they're networked, which is especially relevant when looking at communication infrastructures. These maps reveal how different parts of the world are deeply entangled—economically, politically, ecologically—even if the systems they operate under are vastly different. So yes, mapping is important. It helps us grasp the scale and interconnectedness of these infrastructures. But it also comes with the danger of abstraction. The same clarity that allows us to see global connections often hides the embodied, localized, or more-than-human experiences behind them. These maps might show a cable landing on a stretch of coastline, but they don't show how it touches the ground—what community it intersects with or what ecological systems it might disrupt. That's why I think it's essential to ask: What are we actually seeing when we "see" these maps? And just as importantly, what are we not seeing? Whose stories are being excluded or flattened into symbols and lines?

As architects, designers, and researchers, we do have a certain agency here. We can decide what kinds of maps to produce—what to emphasize, what to make visible, and how to visualize complexity without simplifying it to the point of erasure. That might mean combining cartographic methods with drawing, storytelling, video, or oral histories. It means resisting the tendency to render everything into data points and instead staying attentive to the textures, the voices, and the resistances that don't easily fit on a map. In that sense, the map becomes not just a tool of representation, but of inquiry—and potentially, of accountability.

Global map illustration by Virginia Zangs revealing the physical infrastructure behind our digital world. Undersea cables (white lines) and data centers (neon blue dots) form the backbone of global telecommunications, 2025.

When we started working with these maps, looking at the histories, materialities, and geopolitical implications of the cables, one of the most frustrating things was to find current information. This infrastructure is designed to be invisible to the public. This raises questions about its economic and political interest. For example, who installs the cables, and who profits from this enterprise? Why are certain cables public, while others are private? Where are they made, and what are their global ecological impacts?

Arrival of the Tuvalu Vaka Cable at Funafuti, Tuvalu. Vaka is the first international telecommunications cable to connect Tuvalu, 2025.

Recently, I've been trying to understand how we see what we see in these maps. You're right: these underwater cables lie in the depths of the ocean, out of sight. So when we encounter a planetary map showing these networks, what technologies and systems have made that image possible? Most of the time, we're relying on satellite imagery or on data provided by the very ships that lay the cables. These representations are mediated by infrastructures—often commercial and inaccessible—that shape what we can see and know. It's essential to understand how power and infrastructure mediate our perception of the world.

Then there's the broader question of infrastructure and invisibility. What we typically demand from infrastructure is seamlessness and efficiency. Most people don't want to think about what happens to their waste or how their data travels. Architecture, except for certain moments in history—like in the 1960s and 1970s, when infrastructure sometimes became an aesthetic gesture—has often worked to conceal these systems. Pipes, tubes, cables: they're all often hidden behind walls. This logic extends from buildings to cities to planetary-scale systems.

Laying a cable between continents is an extraordinarily expensive operation, and only a handful of entities—such as Google—can afford to do it. In Tuvalu, for instance—a Pacific island nation with a population of around 10,000, already suffering the effects of sea level rise and actively preparing for the worst-case scenario by proposing the creation of a digital nation—the state could not finance the construction of a subsea cable on its own. Google was already building a cable in that region of the Pacific, and the Tuvaluan government negotiated a branching connection. The project received financial support from Australia. When I traveled to Tuvalu last week, for the first time, I saw a cable landing point on a beach—this surreal moment where a line emerging from the ocean's depths was set to connect Tuvalu to the rest of the world. Until now, the island's communication infrastructure has depended largely on satellite systems like Starlink. There's enormous hope and expectation attached to the cable: it promises faster communication, smoother transactions, and a more integrated position in the digital economy.

But this is a public-private project. And the case of Tuvalu is emblematic of ongoing colonial relations embedded in contemporary infrastructure. Because of the vast costs involved, many nations can only gain access to global communication infrastructure through partnerships with powerful corporations—partnerships that often come with conditions, reinforcing asymmetrical power dynamics in exchange for essential service. The internet, in this sense, is not a commons. It is an extractive, geopolitical technology—structured by inequality and sustained by opacity.

> This infrastructure is vulnerable to external threats like natural erosion, component failure, fishing, and ship anchoring. Deliberate interference, like the recent targeted damage to submarine cables in the Baltic Sea, is also rising, resulting in significant data and financial losses.

Tuvaluans are acutely aware of their reliance on a mix of public and private enterprises, corporations, and governments for basic infrastructure and connectivity. At the same time, they are stewards of traditional knowledge built on centuries of observation and attunement. Each family holds a secret—forms of sensing and predicting environmental changes rooted in deep ecological understanding. Today, the Tuvaluan government is working to integrate this ancestral knowledge with data drawn from digital infrastructures.

Concerns about the vulnerability of internet cables are being addressed through unexpected alternatives: a combination of ancestral knowledge and both local and international infrastructures, including satellite networks from providers such as Starlink, and even proposals for a decentralized network of data centers hosted in Tuvaluan embassies around the world.

> As you said, people usually perceive infrastructure through a notion of efficiency and functionality. For the production of the first transatlantic cable, gutta-percha was used as a material to coat this cable, as it could sustain salt water. The expansion of cable networks coincided with the extinction of the gutta-percha plant in Southeast Asia, because the whole population of that plant was used for the production of these cables. So, this is one

way these colonial implications of extractive practices can be made very explicitly visible. The *khipu* is another good case to address in terms of linking these material artifacts and infrastructures to their colonial reading.

The *khipu* offers a powerful counterpoint. It's another artifact that helps link material infrastructures to their colonial entanglements, but from a completely different perspective—one rooted in embodied knowledge, cultural continuity, and resistance. The first *khipu* I saw was in the Archeological Museum in Santiago, Chile. I was in awe of its beauty, complexity, and sophistication. And I think of it in two ways, particularly in relation to infrastructure.

First, if we place the *khipu* in the genealogy of calculating or information storage devices—like today's computers and data centers—it becomes clear that it wasn't just a carrier of information, but an abstract and interactive medium. The Khipu Kamayuqs, those who created and read *khipus,* used multiple senses to encode and retrieve knowledge from them. They didn't just read them with their eyes; they touched them. The texture of the fibers, the knots, their direction, the colors—all of these elements encoded data. But to access that data, you had to engage with the *khipu* physically. There was a transfer of knowledge through the body, and that knowledge was coproduced by the Khipu Kamayuq, the *khipu* and the ecologies it related to.

This made it extremely difficult for colonial officials to understand or control the system. They needed not only to decode a symbolic language but to engage in cultural and embodied practices they didn't share. That opacity made the *khipus* deeply threatening. They were suspected of harboring secret messages, of enabling rebellion. At one point, they were even used to resist the practice of Catholic confession, which was being imposed on the Indigenous populations. And so, *khipus* were ultimately banned and destroyed. If we place this in a broader conversation about cartographies—about maps, navigation charts, and the forms of abstraction developed by European colonial powers—we can see how mapping itself was a tool for conquest. It used abstraction to render land and sea into navigable, knowable, and therefore ownable forms. In contrast, something like the *khipu* proposes a different way of storing and transmitting knowledge—one that is relational, embodied, and context-specific.

Today, there are ongoing efforts to decode *khipus,* some using artificial intelligence. But in my view, the *khipu* cannot be fully decoded. You might get close to understanding some of the symbolic systems, but without the bodies that once enacted and interpreted that knowledge—bodies erased by colonial violence—you lose the system's core logic. These contemporary decoding efforts risk repeating the same colonial gesture: imposing Western rationalities on a knowledge system that fundamentally transcends them. It's a form of epistemic extraction. The *khipu* reminds us that knowledge doesn't always live in a server, on a screen, or in an externalized storing device. It can be tactile, intimate, and bound to specific bodies, environments, and practices. And that's something that today's infrastructures, with their focus on efficiency and abstraction, often fail to acknowledge.

> In the exhibition, we are discussing some of the critical raw materials that are essential to the construction of data infrastructure. The demand for rare earth metals, copper, cobalt, and lithium, to name a few, is increasing. In addition, the recent development of AI exacerbates extractive practices, justified by the need for endless data accumulation. This demand for expansion and extraction is directly linked to keeping up with our Western, capitalist logic of continuous economic growth. Can we challenge what the "green" and "digital" transition is putting forward as an inevitable necessity?

One of the reasons we've focused on the relationship between these elements and their sites of extraction is because digital infrastructure—and renewable energy as well—have long been presented as "green" or "clean" alternatives to fossil fuels. I absolutely agree that we must move away

The MCHAP 0780 is the largest *khipu* on display at the Museo Chileno de Arte Precolombino in Santiago, Chile, dating from ca. 1500. Composed of 586 camelid knotted fiber cords organized into eight sections of ten sets with up to thirteen sub-levels of information, it holds more than 15,000 items of data.

Aerial view of lithium fields in the Atacama Desert in Chile, 2023.

from fossil fuel dependency. However, what's misleading is the reluctance to acknowledge the destructive and polluting practices that are also embedded in the renewable economy. Electric vehicles are certainly preferable to combustion engines, but they, too, carry a significant environmental cost. Many of us were misled by the promises of the so-called Green New Deal, which suggested we could transition to a new energy regime without addressing its extractive foundations or adapting the way of living in the Global North. This doesn't only apply to digital infrastructures and renewable energies, but also to architecture and design. It's a question of accountability—of understanding where the materials we work with come from, and what the broader implications of using them are.

And among all the elements you mention—rare earths, copper, cobalt—I have been focused on lithium. For me, it was the convergence of two issues. One is its role in batteries and renewable energy. The other is its connection to mental health. There's a strange mirroring in the idea that we need more lithium to sustain our current way of life, while that very lifestyle is producing burnout, depression, and mood disorders that lithium is then prescribed to treat. That double agency and dependency was deeply compelling. We began focusing on the Atacama Desert because it's such a central image in the imaginary of lithium extraction—these vast, otherworldly evaporation pools in the desert, whose aesthetics can be strangely sublime. But behind that image lies a violent chemical process that depletes the land, leaving Indigenous communities—and the nonhuman beings connected to the *salares*—without water.

> Right now in Europe, many sites for potential lithium mines are being proposed in places like Portugal, Germany, and Serbia. This would certainly cause ecological harm and impact local communities. What would the role of the European Union be in managing these scenarios? Can we approach this differently from how it's currently done in Chile?

I hope so. But as things stand, mining legislation in places like Portugal doesn't offer more guarantees than in Chile. It's true—Chile has a long, complex history with extractive industries like copper and lithium. Over time, that's created a strong infrastructure of activism, legal tools, and environmental oversight. There's a network of biologists, lawyers, and community organizations who've been working for years to resist or regulate extractive practices. There have even been court rulings affirming the rights of Indigenous communities to block lithium projects in their territories or requiring proper consultation before extraction proceeds. And the Chilean government has started to move toward establishing a national lithium company and outlining areas where extraction should not be allowed.

That kind of framework doesn't really exist in Europe—at least not yet. While some regions like northern Portugal or parts of Spain have a mining history—like the extraction of wolfram for Germany during World War II—mining hasn't been as prominent in recent decades. The legal structures in many of these regions don't adequately protect local communities or ecosystems. Often, if the European Union classifies an activity as essential for the strategic autonomy of the continent, and the national government agrees, it becomes incredibly difficult to oppose

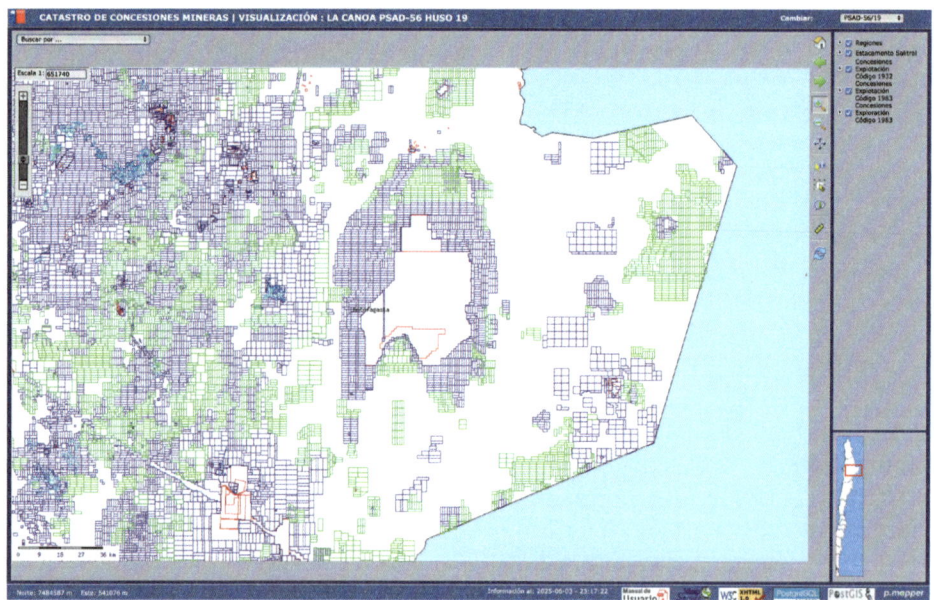

Mining concessions in the Salar de Atacama. Screenshot from Catastro de Concesiones Mineras.

it through legal means. The process is opaque, and participation is limited. That's why we've been working to connect communities in Europe with counterparts in Latin America—especially in Chile—who've been resisting extractivism for decades. We're learning from their legal strategies, their organizing tactics, and the ways they've managed to challenge powerful interests. There are striking parallels—like the need to prove how mining affects local water systems or biodiversity. Across regions like Serbia, Portugal, and Galicia, there's an active exchange of knowledge. We're trying to understand how to resist policies imposed from the highest levels of European bureaucracy, while also acknowledging the motivations behind them. The goal of energy independence is understandable in this global context—but right now, European mining laws often offer very little protection to local environments and populations, and that's something we urgently need to address.

> Do you think there is a potential for developing a sustainable European regulatory framework?

I still believe Europe has the capacity to rethink its approach—to imagine different models for digital infrastructure and alternative notions of progress. Within these global geopolitical shifts, there's an opportunity for Europe to take a distinctive stance on the future of AI, energy, and technology. I hope that possibility is taken seriously. At the same time, Europe's colonial history—and the persistent North–South imbalances within the continent itself—call for caution. These systems of exploitation are deeply rooted and have a way of resurfacing under new guises. For instance, the lithium that mining companies hope to extract from northern Portugal will be exported to other territories, increasingly further north, such as Germany, where it will be processed in a refinery near Bitterfeld-Wolfen, and later used as a key component in electric car batteries. When we ask the director of the refinery, part of the AMG Critical Materials Group, why lithium isn't extracted in Germany, he responds confidently: "It's too polluting." I can't help but wonder: Why replicate the same systems? Why not seize this moment as an opportunity to rethink our approach altogether?

Submarine Cables:
Colonial Roots of Global Communication

Standing glass with contents labeled "Gutta Percha nativ."
Material: gutta-percha; **Size:** 135 × 65 mm; **Date:** 1923; **Collection:** Deutsches Museum; **Signature:** 2019-163

Replica of gutta-percha press by Werner von Siemens.
Material: copper alloy, iron alloy, partially lacquered, wood handle; **Size:** 250 × 420 × 1240 mm; **Manufacturer:** Siemens & Halske; **Date:** 1905; **Collection:** Deutsches Museum; **Signature:** 2459

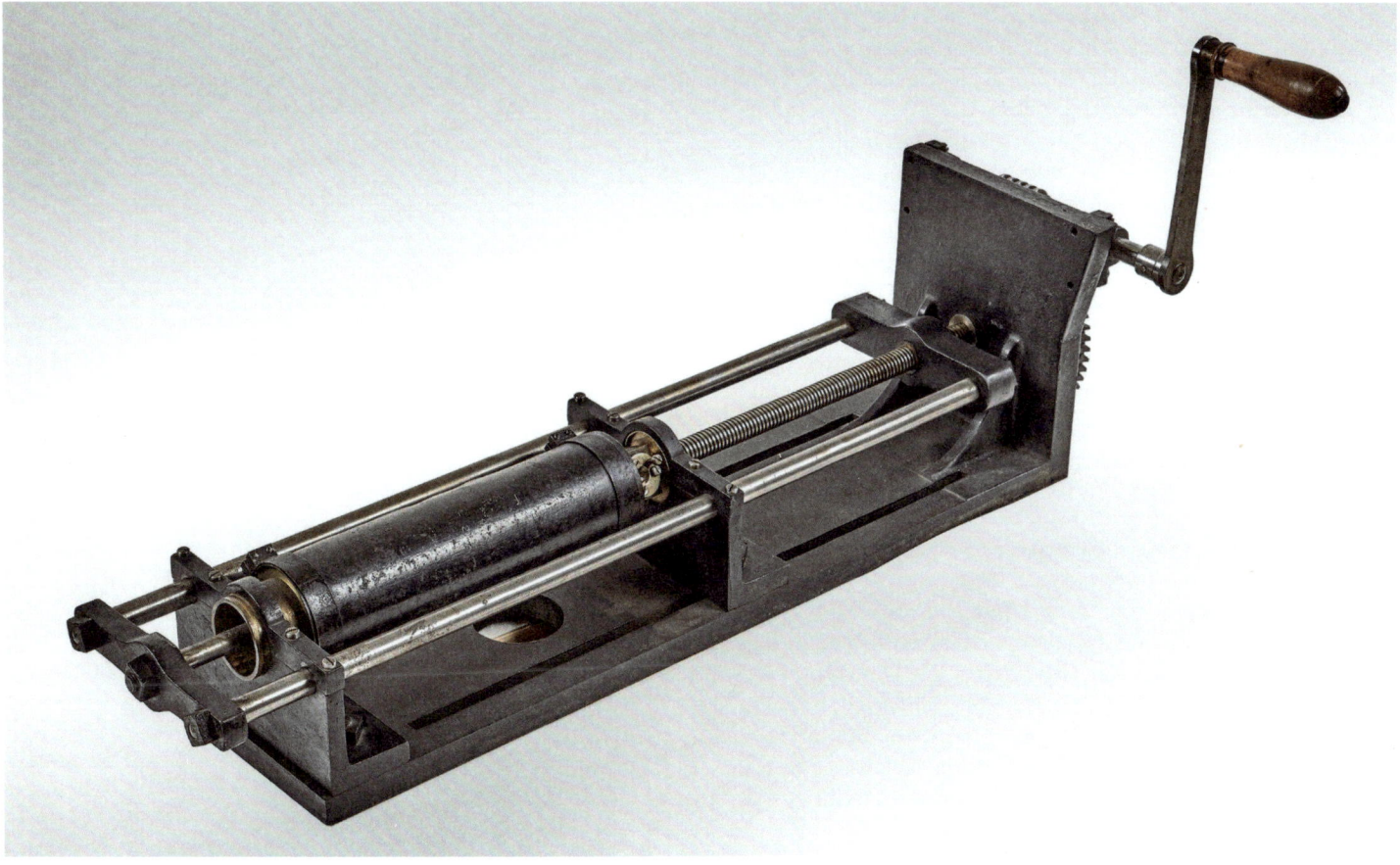

Elemental

Submarine Cables:
Colonial Roots of Global Communication

In 1850, the first submarine cable was laid between Dover in the United Kingdom and Calais in France—a critical endeavor that was severed just a few hours later by a fisherman's anchor. However, a year later, a second attempt was successful, thanks to a more effective coating material made from the gutta-percha plant. This marked the beginning of the global submarine cable industry. The production process was complex and required specifically designed ships, such as the *Faraday I*, launched by the independent cable maker and contractor company Siemens & Halske in 1874.

Beyond the well-known first transatlantic submarine cable, which connected North America and Europe, additional routes soon followed. In 1867, a connection was established between London and Calcutta in India. From the 1880s onward, European cities were connected to Argentina, Australia, South Africa, and Southeast Asia. Fast communication via telegraphy became essential for economic interests, but it also enabled European colonial powers to administer and control their colonies more efficiently, allowing rapid responses to uprisings and armed conflicts.

By the early twentieth century, approximately 370,000 kilometers of underwater cable had been laid. This required an estimated 27,000 tons of gutta-percha (used for the coating of these cables), mostly extracted through colonial plunder from forests on the Malaysian peninsula and the island of Borneo. A single felled gutta-percha tree yielded, on average, 312 grams of this material. By 1883, when the British banned tree felling, gutta-percha had already become extinct on the Malaysian peninsula. Various colonial powers hence sent expeditions to explore the feasibility of extracting this valuable material elsewhere. For example, between 1907 and 1909, the German Empire dispatched an expedition, led by botanist Dr. Rudolf Schlechter, to its colonies in Papua New Guinea. Although gutta-percha was found there, it lacked the quality needed for successful exploitation.

The felling of around eighty-eight million trees for gutta-percha production initiated the rampant exploitation of Southeast Asia's rainforests—a practice that continues today.[1] Traces of colonialism and imperialism remain visible in contemporary maps of global fiber-optic cable networks, which often follow routes established during colonial times. Formerly colonized countries and their populations are still frequently treated as resources to be exploited rather than partners connected on their own terms. Today, the flow of power and capital has shifted from former empires to Big Tech companies such as Amazon, Google, Meta, and Microsoft.[2] These large-scale infrastructure projects grant their owners unprecedented technical and operational control over data, creating a need for administrative and legal authority comparable to that of the state.[3]

This photo essay compiles objects, maps, and archival photographs from various archives and collections, including the Deutsches Museum, the Bildarchiv der Deutschen Kolonialgesellschaft, the Royal Museums Greenwich, and the Library of Congress. It concludes with a series of photographs by artist Trevor Paglen documenting underwater research near transatlantic fiber-optic cable landing sites in the North Pacific and Atlantic Oceans.

---

[1] The text and photo essay draws on and is inspired by the archival investigation of Sara Müller in the collection of the Deutsches Museum. Sara Müller, "Germany Calling! – Tiefseekabel aus Guttapercha," *Deutsches Museum Blog*, June 5, 2024, https://blog.deutsches-museum.de/2034/06/05/germany-calling-tiefseekabel-aus-guttapercha.

[2] Maximilian Jung, "Digital Capitalism Is a Mine Not a Cloud: Exploring the Extractivism at the Root of the Data Economy," *Transnational Institute*, https://www.tni.org/es/articulo/el-capitalismo-digital-es-una-mina-no-una-nube?translation=en.

[3] Keller Easterling, *Extrastatecraft: The Power of Infrastructure Space* (London: Verso, 2014), p. 15.

Extraction of gutta-percha in New Guinea after the tree was felled. The tree's milk could be tapped by making several incisions in the bark.
Medium: diapositive (glass plate);
Size: 85 × 100 mm; **Collection:** Bildarchiv der Deutschen Kolonialgesellschaft, Universitätsbibliothek Frankfurt am Main;
Signature: 025-0289-28

In the German colony of German New Guinea, now part of Papua New Guinea, two men boil the milk of a gutta-percha tree.
Medium: photograph; **Author:** Rudolf Schlechter; **Date:** 1907–1909; **Book:** *The Guttapercha and Rubber Expedition of the Colonial Economic Committee, Economic Committee of the German Colonial Society to Kaiser Wilhelmsland 1907–1909* (Berlin, 1911); **Collection:** Institut für Geographie der Universität Hamburg; **Signature:** PazR 130/3

Elemental

The route of the *Guttapercha and Rubber Expedition of the Colonial Economic Committee, Economic Committee of the German Colonial Society to Kaiser Wilhelmsland*.
Medium: map; **Size:** Scale 1:75,000; **Author:** Rudolf Schlechter, Wilhelm Wernicke; **Date:** 1907–1909; **Book:** *Die Guttapercha- und Kautschuk-Expedition des Kolonial-Wirtschaftlichen Komitees, wirtschaftlicher Ausschuss der Deutschen Kolonialgesellschaft nach Kaiser Wilhelmsland 1907–1909* (Berlin, 1911); **Collection:** Institut für Geographie der Universität Hamburg; **Signature:** PazR 130/3

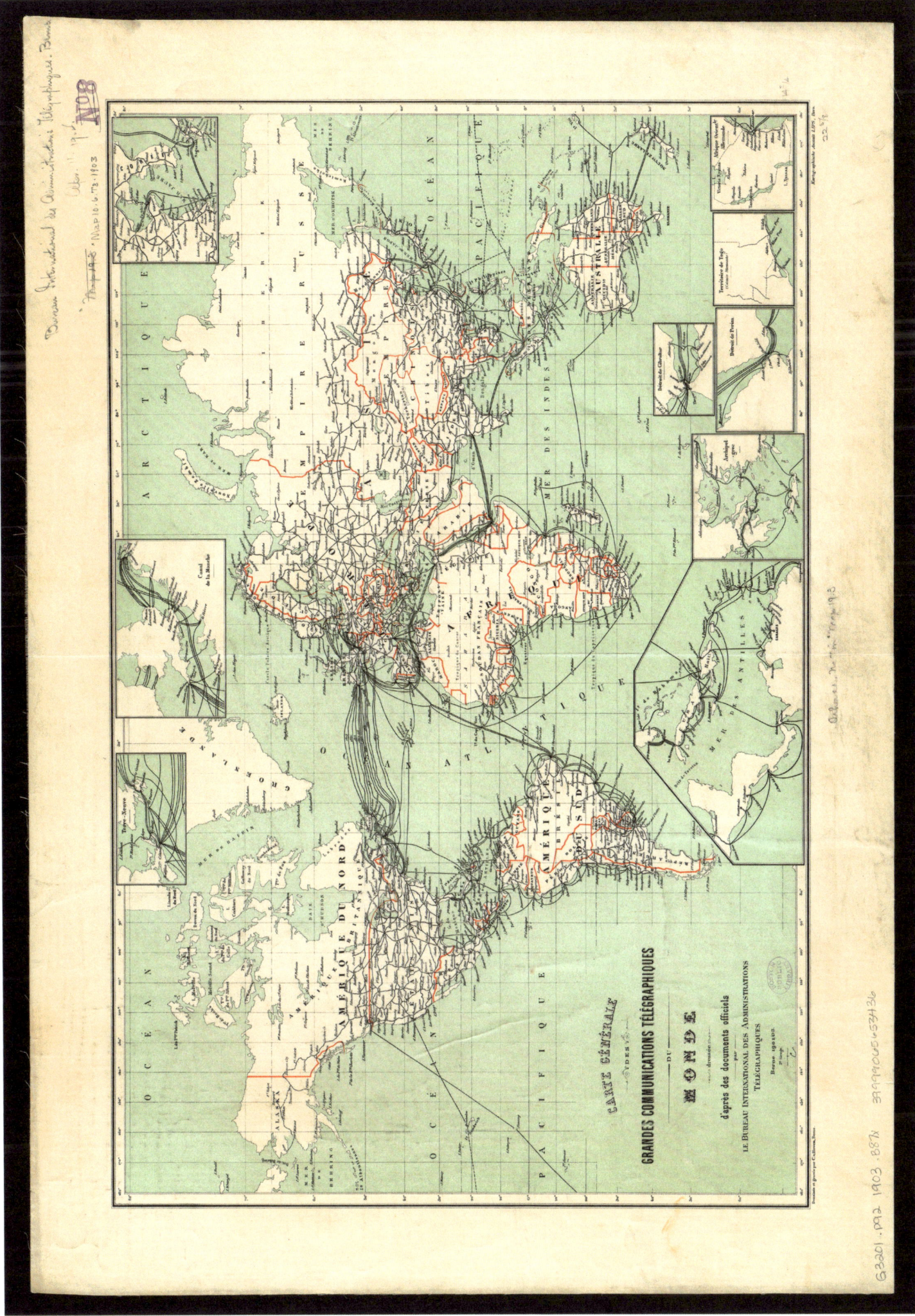

Global telegraph communications.
Medium: map; **Size:** scale ca. 1:70,000,000; **Author:** C. v. Hoven; **Publisher:** International Telegraph Bureau (Bern, 1901)

The Route of the Atlantic Telegraph by Capt. H. Clark, R.N. The Great Eastern: Section of the Bed of the Atlantic. At the top of this print is a map of the route of the cable from Valentia Island in Ireland to Trinity Bay in Newfoundland. At the bottom is a cross-section of the *Great Eastern* showing cable-laying equipment, apartments for people, the steam engine, funnels and coal, horses, and other animals. **Medium:** colored etching; **Size:** 386 × 365 mm; **Artist:** Captain H. Clark; **Date:** July 20, 1865; **Publisher:** Stevens; **Collection:** National Maritime Museum, Greenwich, London; **Signature:** PAG8266

Oil painting of the first transatlantic cable installation to North America by the Siemens Brothers cable steamer *Faraday* in 1874.
Medium: oil on canvas; **Size:** 2,700 × 1,500 mm; **Artist:** Claus Bergen, Munich; **Date:** 1932; **Collection:** Deutsches Museum; **Signature:** 66050

The first cable ship *Faraday* built by Siemens & Halske, set out on its maiden transatlantic crossing. Its distinguishing features included two paddle wheels on the sides and an additional rudder on the bow, which made the ship highly maneuverable. Substructures on the deck also enabled the cable to be laid from either the bow or the stern.
Medium: photograph; **Date:** 1874; **Collection:** Siemens Historical Institute

Elemental

The deep-sea cable inside the *Faraday II* ship during the laying of an insulated gutta-percha cable between England and France.
Medium: photograph; **Date:** May 1910; **Collection:** Siemens Historical Institute

Manufacture of the Atlantic telegraph cable. Machines covering the cable wire with gutta-percha at the Gutta Percha Company's works, Wharf Road in Greenwich.
Medium: wood engraving; **Size:** 233 × 162 mm; **Date:** 1857; **Publisher:** Illustrated London News

The Eighth Wonder of the World: The Atlantic Cable. Allegorical scene showing Neptune with a trident in foreground, and lion representing Great Britain holding one end of the Atlantic cable and eagle representing the United States holding the other end of the cable, with ocean between them and cities behind them. Includes portrait of the inventor, Cyrus Field, at top center.
Medium: color lithograph; **Date:** 1866; **Publisher:** Kimmel & Forster, NY; **Collection:** Library of Congress; **Signature:** ds 04508

Section of first transatlantic submarine cable between Europe and the United States with its jewelry box covered in stylized French lilies.
Material: metal, gutta-percha; **Size:** 118 × 12 mm; **Manufacturer:** Atlantic Telegraph; **Date:** 1858; **Collection:** Deutsches Museum; **Signature:** 63436T1

A sample longitudinal section of a special cliff cable for the Atlantic cable.
Material: metal, gutta-percha, textile; **Size:** 505 × 65 mm; **Manufacturer:** Siemens Brothers & Co.; **Date:** 1894; **Collection:** Deutsches Museum; **Signature:** 19813

A sample cross section of the Atlantic telegraph cable between Waterville, Ireland, and Canso, Canada.
Material: gutta-percha, jute, brass; **Size:** 24 × 67 mm; **Manufacturer:** Siemens Brothers & Co.; **Date:** 1885; **Collection:** Deutsches Museum; **Signature:** 19812

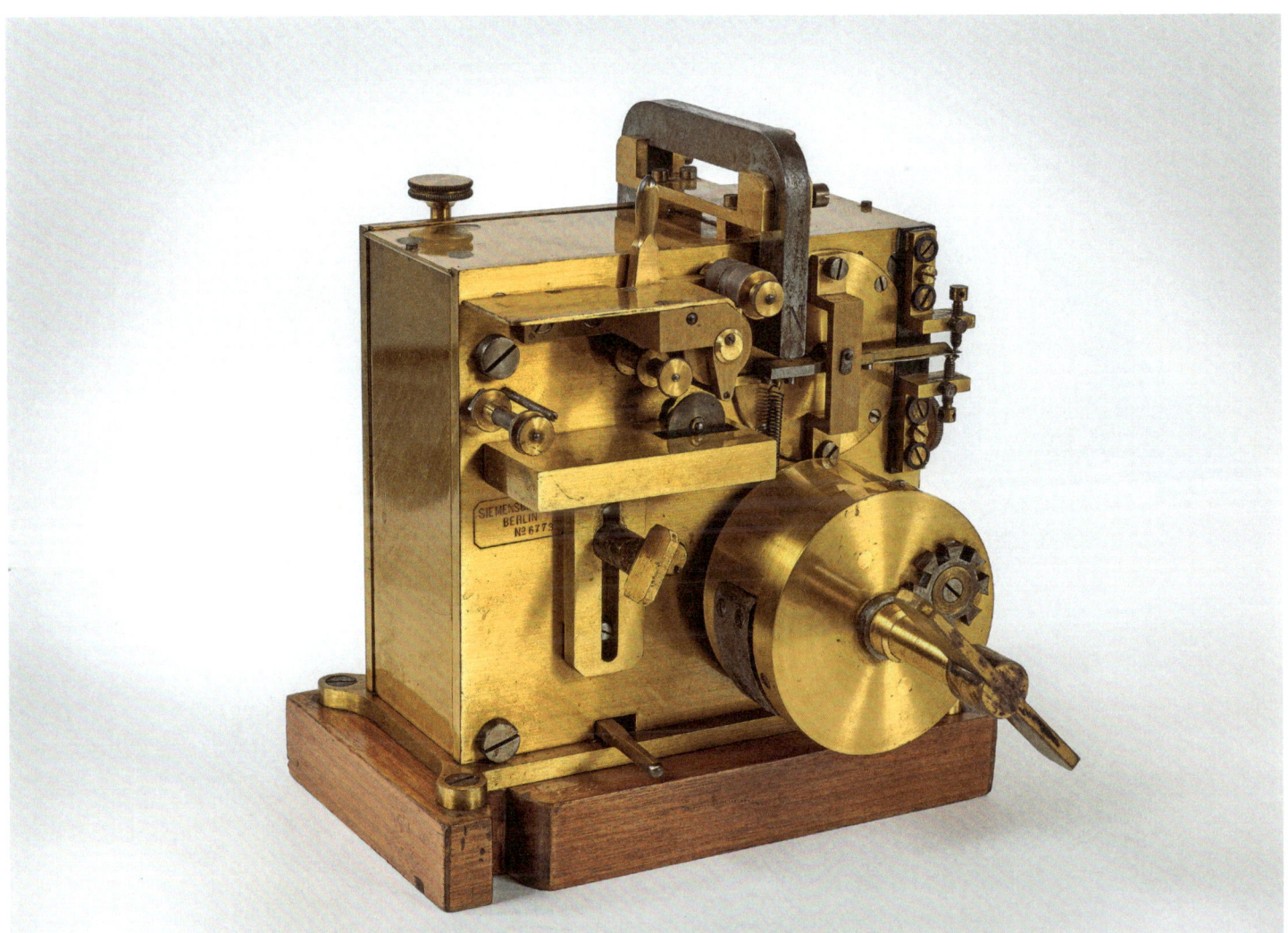

A spring-driven receiver (polarized color recorder) for the Indo-European automatic telegraph system.
Material: metal, wood; **Size:** 210 × 115.5 × 190 mm; **Manufacturer:** Siemens & Halske; **Date:** 1867; **Collection:** Deutsches Museum; **Signature:** 2390

Effect of the Submarine Telegraph; or, Peace and Goodwill between England and France. Two angelic figures, with wings carrying an olive branch and holding hands, following the submarine telegraph cable across the bottom of the English Channel between Dover, England, and Calais, France.
Medium: print, wood engraving; **Size:** 207 × 276 mm; **Date:** September 14, 1850; **Collection:** Library of Congress; **Signature:** pga 14068

Columbus III
NSA/GCHQ-Tapped Undersea Cable; Atlantic Ocean, 2015; C-print; 48 × 60 in.; Trevor Paglen

Bahamas Internet Cable System (BICS-1)
NSA/GCHQ-Tapped Undersea Cable; Atlantic Ocean, 2015; C-print; 60 × 48 in.; Trevor Paglen

NSA-Tapped Undersea Cables
North Pacific Ocean, 2016; C-print; 48 × 72 in.; Trevor Paglen

The *Necro*politics of Big AI

*Necro*political Tricks and "Death Worlds"

In the early 2000s, we were being sold "the cloud" as an immaterial, modern alternative to analogue storage for the digital world. We moved from storing documents, music, and photos on our personal computer hard drives to the hard drives of remote storage companies. Today, we're similarly being sold AI in a way that maintains a level of disconnect to the internet's materialities and impacts. The end goal this time is to make content into data that can be repurposed quickly and cheaply. AI is made of accumulated data in the cloud. It is now seemingly everywhere, but AI isn't new.

Large language models (LLM) and machine learning (ML) technologies have been around since the 1950s, and ML based on artificial neural networks since the 2010s. This is important because these innovations laid the groundwork for what we now commonly understand as "generative AI" (genAI), which has exploded in popularity since 2022 because of products like ChatGPT, DALL-E, Midjourney, Copilot, and others. These AI products continue to multiply; they're being rapidly embedded into existing enterprise software, text-editing applications, and popular search engines—foisted upon users as a new default setting. It seems like all content, all data, is up for grabs for AI ends. Often these (ultimate) "ends" are framed as "AI agents" toward "artificial general intelligence," or AGI.[1] However, in order to avoid confusion with this marketing hype of AGI, I'll use "Big AI" to describe the major multinational cloud service providers that dominate the market and own the computational infrastructure for AI. Defining AI as Big AI becomes important. It enables conversations across disciplines and territories. As Ali Alkhatib explains it:

Defining AI along political and ideological language allows us to think about things we experience and recognize productively as AI, without needing the self-serving supervision of computer scientists to allow or direct our collective work. We can recognize, based on our own knowledge and experience as people who deal with these systems, what's part of this overarching project of disempowerment by the way that it renders autonomy farther away from us, by the way that it alienates our authority on the subjects of our own expertise.[2]

The idea here is that AI becomes a technological domain guarded by those (CEOs, engineers, computer scientists) who build its systems, train its models, and sell its visions. But Big AI is much more than this, and, as Abeba Birhane argues, "We should shed the idea that AI is a technological artifact with political features and recognize it as a political artifact through and through."[3]

So, then, what are AI's—and, specifically—Big AI's politics? And how do these politics materialize? Who benefits and who pays? What are the costs and consequences? How are these politics activated?

Most notably, Big AI *succeeds by failing* to admit its limitations—*deception* and *distraction* as tactics. For example, Eryk Salvaggio writes: "Distracting politicians with fears of sentience . . . is an excellent distraction from meaningful policy and regulatory frameworks that might constrain those profits."[4] So, Big AI *succeeds* by *deceiving* users in framing AI as either near-sentient or as having emergent qualities, or as capable of meaning-making or even hallucinating. Big AI *succeeds* by *downplaying* its operating costs and lack of business model. And in addition to questioning the very essence and viability of Big AI, critical scholars, journalists, artists, and activists have also put into question both the perceived inevitability of AI as a new necessary technology to run the world, and the capacity for the planet (finite) to sustain AI data center projects (exponential growth). This environmental critique is in line with, but also in addition to, other harms identified: that Big AI creates tools of mass plagiarism, deskills workers, bullshits as its core function, undoes critical thinking and empathy in health and educational realms, extracts from artists and authors without consent or compensation, amplifies carceral logics and fascist aesthetics,[5] and relies on outsourced human labor to label video, images, and text in a way that reifies class-based oppression in the rural United States and colonial violence across Asia and Africa, by the US and China, as emotional sacrifice zones meant to absorb the trauma of often vile internet content used to train AI.[6] And so on. Fascism—or, what political analysts Naomi Klein and Astra Taylor specify as "end times fascism"—protects AI because AI is a project of empire building that is itself dependent on resource extraction.[7] As Jasmina Tacheva and Srividya Ramasubramanian explain it, "this networked and distributed global order is rooted in heteropatriarchy, racial capitalism, white supremacy, and coloniality and perpetuates its influence through the mechanisms of extractivism, automation, essentialism, surveillance, and containment."[8]

AI companies present current harms and failures as temporary stepping stones toward a greater future. This temporal framing is crucial: the public is asked to be patient and told that AI's potential is primarily limited by insufficient power resources.[9] Once we provide enough electricity, their CEOs claim, AI will deliver on its promised benefits (though generally grand, those benefits remain opaque). As James Bridle explains it, "Driven by the logic of contemporary capitalism and the energy requirements of computation itself, the deepest need of an AI in the present era is the fuel for its own expansion."[10] That expansion is both infrastructural and ideological.

On our already climate-stressed planet, however, this ideological narrative is a deceptive (stupid) political fiction disguised as progress. But, as tech journalist Brian Merchant makes clear, "Stupid or not, it's a powerful fiction."[11] What we're witnessing is the emergence of a technology-dominated future where climate concerns are ignored, and natural resources are exploited, primarily to power AI systems for profit, as Big AI itself struggles to break even.

My (counter)proposal, then, is that we be specific about the politics of Big AI. One way to do this is to make clear that Big AI is not just political; it's *necro*political, as in, a maker of "death worlds," a concept I borrow from Achille Mbembe to explain environments where certain populations are subjected to conditions that threaten their existence or livelihood through perpetual precarity, social death, and physical endangerment while being excluded from legal protections and political recognition.[12] Applied to contemporary AI systems, the concept of death worlds illuminates how Big AI increasingly wields algorithmic sovereignty to sort populations into those who benefit from technological advancement and those subjected to its extractive and harmful aspects.

Seen as death worlds, it becomes harder to determine what separates planetary from human impacts—they are, in fact, completely intertwined and utterly inseparable within a forceful critique of Big AI. Put more simply: we must understand the politics (as *necro*politics) of Big AI in order to think through its environmental harms as always already extractive and carceral logics.

Stupid, but Costly

Big AI *is* enormously costly—and barely profitable (beyond the marketing of its potential). Advanced AI systems require immense processing capacity requiring thousands of graphics processing units (GPUs) to train and run models. This is why Big AI data center energy costs are largely associated with running computing infrastructure, and why Big AI fails to make a profit—the demands exceed the returns. Because of the massive energy demands in particular, Big AI stresses existing public power grids and increasingly diverts water away from people and farms—used to cool data servers instead.[13]

In March 2025, the *New York Times* released a video explainer titled "How A.I. Companies Are Turning into Energy Companies" to show just how entangled and codependent the AI and energy industries have become—morphing into one another.[14] While that video fails to go into climate change or the environmental impacts of AI data centers at this scale (current and imagined for the future), Merchant described that relationship, between these industries, as "automating climate change"—evident to him even back in 2019, pre–Big AI explosion into the mainstream.[15] Big AI's (now) urgent need for energy has reinvigorated fossil fuel, coal, and nuclear industries while also, simultaneously, touting a green (more efficient) turn by way of renewables, like solar and wind.

Despite these (often rhetorical) twists and turns to sell the public on a better and brighter AI-enabled future, researchers have calculated that air pollution "derived from the huge amounts of energy needed to run data centres has been linked to treating cancers, asthma and other related issues."[16] Researchers also found that the cost of treating illnesses connected to AI data-center pollution was already valued at $1.5 billion (in 2023), and had only increased since.[17] These are planetary diagnoses largely (willfully) ignored by Big AI. They're ignored because the *necro*politics of Big AI are driven by the idea of profit and the ideal of endless growth at the (very real, very material and ideological) expense of earth and its inhabitants.

Misguided pleas for profit and growth are being illustrated in real time today.[18] As we've seen revealed in the past few months, Big AI is owned by companies in line with "dark enlightenment," which is an antidemocratic and neoreactionary movement that is explicitly *against* values of public health, equity, diversity, inclusion, and accessibility as well as being against environmentalism.[19] In this sense, Big AI is overtly and ambitiously *necro*political, seeped in visions informed by TESCREAL politics,[20] or what Corey Pein has called "total corporate despotism."[21] This is also why we're seeing the US government scrub posts critical of Amazon, Microsoft, and AI companies from the Federal Trade Commission website, the thrust there being techno-fascism as authoritarianism driven by technocrats.[22] The hard-right shift of the US government today is manifesting as what could very well be unlawful "deportations" and detentions at the border, mega-prisons like those in El Salvador, attacks on trans people's rights, and the policing of women's bodies—much of which is enabled by data technologies themselves and funding from Big Tech—and is reminiscent of early fascism. At this juncture, AI is taking too long to deliver any tangible public benefits and is having to trick the public into adopting and deploying it, especially as it becomes more obviously a tool of authoritarianism.[23]

Big AI Takeover

There are daily news headlines that illustrate the scale and—inadvertently, maybe—the ridiculousness of the current Big AI takeover. Of note was Sam Altman's statement that OpenAI would require seven trillion dollars to get AI up and running in the US only to be shattered months later by DeepSeek, a Chinese version that purportedly costs much less to run. While there is now debate about the true costs of each project—as these things are notoriously hard to trust, track, or measure—the calculations vary from millions to billions in total server capital expenditure, with most of the spending to maintain and operate GPU clusters. This is why Big AI in the US is on a mission to rebuild and expand its empire by becoming, or making deals with, energy and mining industries—the main ones used to power AI currently are coal and other fossil fuels—while also looking for locations for AI data centers that provide embedded efficiencies, such as low humidity and a cool climate, while also hoping for the annexation of Canada and Greenland, and threatening Ukraine, places perceived to be rich in "rare earth" minerals needed for advancements in AI. Energy and mining companies, in turn, use AI in exploration, drilling, extraction, refining, pipeline management and maintenance, and in modeling and monitoring equipment and so on. To be certain, Big AI, energy, and mineral extraction are deeply imbricated, but their imbrication is also a projection of infrastructural prowess as it paints a picture for the public that these large-scale projects are sophisticated and necessarily futuristic in their complexity.

In Canada, similar aspirations (to the US and China) are being expressed via the federal government's AI Compute Challenge and, for example, with private investor Kevin O'Leary making plans to build a $70 billion data center called Wonder Valley, in Alberta, praised by right-wing politicians like Alberta Premier Danielle Smith. So too the UK, going all in on Big AI: AI "will be 'mainlined into the veins' of the nation, ministers have announced, with a multibillion-pound investment in the UK's computing capacity despite widespread public fear about the technology's effects," explains *The Guardian* in 2025.[24] It's not hyperbole to say that AI is being forced onto and into the general public by the corporate world—with liberal *and* conservative government support. In the UK, support is in the order of "£200 million a day in private investment since July (2024)."[25] To compare, in the past decade, "China's government has invested $912 billion," second only to the US.[26] At this juncture, we're seeing massive investments, but also, recently, some new tentativeness, with, for example, Microsoft recently canceling several data center leases in the US (while nevertheless spending $80 billion on AI infrastructure in this fiscal year).

While there have been important interventions calculating the impacts of AI through metrics—energy, minerals, or water used to train AI models, for example—the argument presented here is that these are not sufficient if we are to properly understand the political economy of Big AI. It's not enough, at this juncture, to simply want more energy-efficient AI models. It's not enough to want less biased datasets. It's not enough, or even possible, to want responsible or ethical AI guardrails. These technical fixes fail to address how AI has fundamentally transformed every aspect of human experience into extractable data commodities, creating inescapable digital architectures that lock us all into AI-mediated worlds where our relationships, knowledge, and ways of being are increasingly determined through datafication and by algorithmic systems.

It's very hard to say if the Big AI bubble will burst, or how and when, exactly, but it remains very clear that AI is not the promised "technofix" for people and the planet. It's also very clear that it doesn't have to "work" in any real sense of the word to nevertheless function as a tool for end times fascism. What is also very clear is that massive investments in AI infrastructure have already demonstrated the harms and the devastation of the *necro*political capitalism in which it basks. In other words, AI investors not only benefit from but actively revel in this harmful system, suggesting that its rapid advancement comes at significant social and environmental costs that are being ignored or even celebrated rather than addressed. The benefits of an AI-data-commodified world—to fascism, totalitarianism, and authoritarianism—are already evident. As Big AI increasingly dominates the landscape, the way we understand and resist AI *now* is crucial for human and planetary *futures*.

1. AGI is a marketing gimmick used by AI companies to promise a greater potential from current AI instantiations. As a result, however, LLM/ML have been subsumed into this marketing language of genAI/AI/AGI. In order to make a distinction between the historical work of AI as LLM/ML (that could be used in other perhaps more local and domain-specific, less extractive ways) versus the current marketing terms of genAI/AI/AGI, I'll refer to the latter cluster as "Big AI" to describe the major multinational cloud service providers that dominate the market and own the computational infrastructure for AI.

2. Ali Alkhatib, "Defining AI," *Ali Alkhatib* (blog), December 6, 2024, https://ali-alkhatib.com/blog/defining-ai.

3. The quote continues: "AI is an ideological project to shift authority and autonomy away from individuals, towards centralized structures of power. Projects that claim to 'democratize' AI routinely conflate 'democratization' with 'commodification.' Even open-source AI projects often borrow from libertarian ideologies to help manufacture little fiefdoms." Abeba Birhane, "Bending the Arc of AI towards the Public Interest," AI Accountability Lab, *Aial.ie,* February 18, 2025, https://aial.ie/pages/aiparis/.

4. Eryk Salvaggio, "Fear of AI Is Profitable," *Cybernetic Forests* (Substack blog), April 2, 2023, https://cyberneticforests.substack.com/p/fear-of-ai-is-profitable.

5. See Roland Meyer (@bildoperationen @tldr.Nettime.Org), "AI Images and Fascist Propaganda," *Tldr.Nettime,* January 10, 2024, https://tldr.nettime.org/@bildoperationen/111730421553799071.

6. Leslie Stahl, Aliza Chasan, Shachar Bar-On, and Jinsol Jung, "Kenyan Workers with AI Jobs Thought They Had Tickets to the Future until the Grim Reality Set In," *CBS News,* November 24, 2024, https://www.cbsnews.com/news/ai-work-kenya-exploitation-60-minutes/.

7. Naomi Klein and Astra Taylor, "The Rise of End Times Fascism," *The Guardian,* April 13, 2025, https://www.theguardian.com/us-news/ng-interactive/2025/apr/13/end-times-fascism-far-right-trump-musk.

8. Jasmina Tacheva and Srividya Ramasubramanian, "AI Empire: Unraveling the Interlocking Systems of Oppression in Generative AI's Global Order," *Big Data & Society* 10, no. 2 (2023): 1–13, https://doi.org/10.1177/20539517231219241.

9. Eryk Salvaggio, "Most Researchers Do Not Believe AGI Is Imminent: Why Do Policymakers Act Otherwise?," *Tech Policy Press,* March 19, 2025, https://techpolicy.press/most-researchers-do-not-believe-agi-is-imminent-why-do-policymakers-act-otherwise.

10. James Bridle, *Ways of Being: Animals, Plants, Machines: The Search for a Planetary Intelligence* (New York: Farrar, Straus and Giroux, 2023), pp. 3–7.

11. Brian Merchant, "What's Really Behind Elon Musk and DOGE's AI Schemes," *Blood in the Machine,* February 25, 2025, https://www.bloodinthemachine.com/p/whats-really-behind-elon-musk-and.

12. Antonio Pele, "Achille Mbembe: Necropolitics," *Critical Legal Thinking,* March 2, 2020, https://criticallegalthinking.com/2020/03/02/achille-mbembe-necropolitics/.

13. Dustin Edwards, *Enduring Digital Damage: Rhetorical Reckonings for Planetary Survival* (Tuscaloosa: University of Alabama Press, 2025).

14. Karen Weise, Laura Bult, James Surdam, and Ramon Dompor, "How A.I. Companies Are Turning into Energy Companies," *The New York Times,* video, March 17, 2025, https://www.nytimes.com/video/business/energy-environment/100000010036088/how-ai-companies-are-turning-into-energy-companies.html.

15. Brian Merchant, "How Google, Microsoft, and Big Tech Are Automating the Climate Crisis," *Gizmodo,* February 21, 2019, https://gizmodo.com/how-google-microsoft-and-big-tech-are-automating-the-1832790799.

16. Cristina Criddle and Stephanie Stacey, "Pollution from Big Tech's Data Centre Boom Costs US Public Health $5.4bn," *Financial Times,* February 23, 2025, https://archive.ph/N0UGX.

17. "Air Pollution and the Public Health Costs of AI," California Institute of Technology, December 10, 2024, https://www.caltech.edu/about/news/air-pollution-and-the-public-health-costs-of-ai; Troy Wolverton, "AI-Induced Pollution Could Cost Billions and Harm Thousands," *San Francisco Examiner,* February 11, 2025, https://www.sfexaminer.com/news/technology/ai-induced-pollution-could-kill-hundreds-cost-billions-researchers-say/article_6449a044-e811-11ef-88d2-473a3ec5a724.html.

18. Naomi Klein and Astra Taylor, "The Rise of End Times Fascism," *The Guardian,* April 13, 2025, https://www.theguardian.com/us-news/ng-interactive/2025/apr/13/end-times-fascism-far-right-trump-musk.

19. Jason Wilson, "He's Anti-Democracy and Pro-Trump: The Obscure 'Dark Enlightenment' Blogger Influencing the Next US Administration," *The Guardian,* US News, December 21, 2024, https://www.theguardian.com/us-news/2024/dec/21/curtis-yarvin-trump; Andrew Prokop, "Who Is Curtis Yarvin, the Monarchist, Anti-Democracy Blogger?" *Vox,* October 24, 2022, https://www.vox.com/policy-and-politics/23373795/curtis-yarvin-neoreaction-redpill-moldbug; Chris Lehmann, "The Reactionary Prophet of Silicon Valley," *The Nation,* October 27, 2022, https://www.thenation.com/article/politics/curtis-yarvin/.

20. Émile P. Torres, "TESCREALism: The Acronym behind Our Wildest AI Dreams and Nightmares," *Truthdig,* June 15, 2023, https://www.truthdig.com/articles/the-acronym-behind-our-wildest-ai-dreams-and-nightmares.

21. Corey Pein, "Mouthbreathing Machiavellis Dream of a Silicon Reich," *The Baffler,* May 19, 2014, https://thebaffler.com/latest/mouthbreathing-machiavellis.

22. Makena Kelly, "FTC Removes Posts Critical of Amazon, Microsoft, and AI Companies," *WIRED,* March 18, 2025, https://www.wired.com/story/federal-trade-commission-removed-blogs-critical-of-ai-amazon-microsoft.

23. E.g., DOGE, or cloud companies' support of the Israeli military. DOGE has done tremendous work dismantling democracy in the US, lining the pockets of the already ultrarich. The collective No Tech for Apartheid called Google's partnership with Israeli start-up Wiz "shameless genocide profiteers." A similar recent protest against Microsoft led to two employees being fired over their concerns that AI was being used for genocide. There's a "whole vibe" to Big AI and its necropolitics. See: Campaign No Tech For Apartheid, "STATEMENT on Google Cloud's Acquisition of Israeli Start-Up Wiz," *Medium*, March 19, 2025, https://medium.com/@notechforapartheid/statement-on-google-clouds-acquisition-of-israeli-start-up-wiz-987cbaf34b0f.

24. Robert Booth, "'Mainlined into UK's Veins:' Labour Announces Huge Public Rollout of AI," *The Guardian,* January 12, 2025, https://www.theguardian.com/politics/2025/jan/12/mainlined-into-uks-veins-labour-announces-huge-public-rollout-of-ai.

25. UK Government, "UK AI Sector Attracts £200 Million a Day in Private Investment since July," tps://www.gov.uk/government/news/uk-ai-sector-attracts-200-million-a-day-in-private-investment-since-july.

26. Stanford Center on China's Economy and Institutions, "Government Venture Capital and AI Development in China," December 1, 2024, https://sccei.fsi.stanford.edu/china-briefs/government-venture-capital-and-ai-development-china; IntelligentCIO, "USA Leading the Charge on AI Investment – Intelligent CIO North America," https://www.intelligentcio.com/north-america/2024/08/08/usa-leading-the-charge-on-ai-investment/.

Lithium Mining in the Atacama Desert

Fragile Land, Finite Salts

The Salar de Atacama salt flat in Chile is one of the few places on Earth where lithium is found in very high concentration. The country produces approximately 79 percent of the European Union's supply of this natural element.[1] Lithium is found in mineral salts suspended in subterranean brine reservoirs, which are pumped into ponds on the surface of the desert and exposed to several cycles of solar evaporation. Around 2.2 million liters of water are evaporated to produce one ton of lithium in the world's driest place.[2]

Catherine Hyland's series of images from 2018 documents this surreal process. The liquid turns increasingly acidic green and yellow as the salt concentration rises—a scene set against the backdrop of a scorching desert and clear blue sky, capturing the region's ongoing drought. The raw material is exported from the Atacama to other parts of the world to fuel the so-called "green energy transition," which in turn powers electric cars and continuously lit screens with lithium-ion batteries.

However, these minerals are neither "lifeless" nor mere "commodities." They are rich with microorganisms native to this desert and are deeply intertwined with its earth—both ecologically and humanly. Images of nearly dried-up riverbeds, deserted villages, and abandoned cars hint at past human inhabitation. Indigenous communities, such as the Atacameño, have been stewards of this land for generations, now relying on the few jobs provided by the mining companies that are slowly destroying their territory and livelihood.[3] While lithium has historically been praised for its medical and mythical ability to keep mind and body sane, Hyland's artistic research reveals the consequences of forcibly extracting this vital element from its fragile land in service of the digital and green transition.

[1] Council of the European Union, "Critical Raw Materials," Consilium, https://www.consilium.europa.eu/en/infographics/critical-raw-materials/.

[2] Maeve Campbell, "In Pictures: South America's 'Lithium Fields' Reveal the Dark Side of Our Electric Future," *Euronews,* February 1, 2022, https://www.euronews.com/green/2022/02/01/south-america-s-lithium-fields-reveal-the-dark-side-of-our-electric-future.

[3] Marina Otero Verzier, "The Knots of the Metaverse: On Data Infrastructures and Indigenous Rights," in *Acid Clouds: Mapping Data Centre Topologies,* eds. Niels Schrader and Jorinde Seijdel (Rotterdam: nai010 Publishers, 2024), p. 157.

Lithium Mining in the Atacama Desert

Lithium Mining in the Atacama Desert

Elemental 59

Lithium Mining in the Atacama Desert

Elemental 63

Elemental

Lithium Mining in the Atacama Desert

Exhuming Earth: Extraction and Resistance in the Age of "Green Transition"

Albemarle lithium mining facilities in the Salar de Atacama, Chile, 2019.

Plundering for the "Green Transition"

A new "gold rush" is underway. Around the world, unique ecosystems, agricultural or pastoral lands, and the ancestral territories of traditional and Indigenous peoples are facing a new wave of resource extraction. This is due to a rapidly growing demand for minerals such as lithium, copper, and cobalt, which are considered central to the "green transition." In theory, the green transition refers to efforts to move economies away from their dependence on fossil fuels—something that is crucial to mitigating the effects of climate change. In practice, however, the "green transition" is turning out to be something quite different.

When we look at global investment in the "green transition," we see that most of the focus has been on individual road vehicles, and the shift from internal combustion engines to electric vehicles (EVs). The prioritizing of investment in EVs might appear strange if we consider how the impact of individual road vehicles on global CO_2 emissions is quite small, corresponding to 7.2 percent of global CO_2 emissions.[1] Ecologically, 7.2 percent is a dramatic number that needs to be brought down. But it does not explain why individual road transportation has become the main focus of climate mitigation efforts, over all other avenues not only of decarbonization but more broadly of ecological reparation. EVs have taken priority over shifting from individual to public transport, home retrofitting, building and material reuse, degrowth, reducing pendular movements, stopping deforestation or shifting agricultural practices, among many others.

Of course, the demand for EVs has grown not out of research-led environmental calculations but out of marketing a new business model—a process in which Tesla played a key role. While climate change mitigation might have been part of the initial push, this should be understood as an expansion of the energy market. For example, if the point was to phase out oil, then we should notice how most scenarios for oil demand by 2050 point toward a continued increase. As much is predicted by OPEC, for whom "in the long term, global oil demand is expected to increase by almost 18 mb/d, rising from 102.2 mb/d in 2023 to 120.1 mb/d in 2050."[2] How is this possible? Exxon Mobil explains: "The large majority of the world's oil is and will be used for industrial processes, such as manufacturing and chemical production, along with heavy-duty transportation like shipping, trucking, and aviation."[3] In other words, the green transition, with its focus on EVs, is not leading to a shift in the capitalist demand for high energy consumption at a planetary level. On the contrary, it evidences how capitalism co-opted a well-intended proposal to mitigate climate change into a political pacifier, promoting the collective sublimation of guilty feelings by driving "clean" vehicles, while maintaining untouched (in fact making much worse) most other drivers of ecological collapse.[4]

Photo of dried lagoons in Peine, Salar de Atacama, 2019.

Exhumations and Fake Geographies

Like other cars, EVs are reliant on materials such as plastics, rubber, and aluminum. But unlike other cars EVs also require a huge amount of minerals for their batteries. Due to the increase in demand for EVs, minerals such as lithium, copper, and cobalt have become some of the most precious commodities in the world. This is visible in the US's and Elon Musk's reported interference in the 2019 coup against the Bolivian president Evo Morales, or in Trump's attempt at brokering "peace" between Ukraine and Russia via guaranteeing access to its mineral deposits—just to offer two examples. Take lithium: 87 percent of all lithium extracted globally is destined for batteries. Of this, 89 percent are car batteries (the remaining 7 percent are for consumer electronics, and 4 percent for stationary storage). Despite common arguments by the mining industry that we all need minerals for use in our mobiles and laptops, without EVs, such acceleration of the extractive frontier would not be taking place.

But it is. In fact, if we look at the Atacama Desert in Chile, home to the Salar de Atacama, one of the world's most important lithium sources, the acceleration in extraction is clearly visible from 2010 onward, in the expansion of US-based Albermarle and Chilean SQM lithium extraction facilities. This is the "lithium triangle," an imaginary geographical area resulting from drawing a straight line across the Andes mountains between the Salar de Atacama in Chile, the Salar de Hombre Muerto in Argentina, and the Salar de Uyuni in Bolivia. This made-up triangle contains the world's largest concentrations of lithium and has become a global attractor of mining investment.

Salares (or salt flats) are dried lake beds with underground reservoirs containing high concentrations of dissolved salts, such as lithium, potassium, and sodium. This is a result of activity in the surrounding volcanoes of the Andes cordillera. Over millions of years, minerals have accumulated in these peaks that descend when the ices melts in spring, leaching into the lower-level soils, and eventually settling in the depressions at the bottom of the *salares*. To extract lithium, exhuming it from the depths of the earth, requires pumping the salt-rich brines from beneath the *salares'* crust into a series of large, shallow evaporation pools. Brine is rich in water while containing only traces of lithium. On average, for each ton of lithium, 500,000 gallons of water are also extracted.[5]

While across the lithium triangle, lithium is extracted from brine (as is in certain areas of the US and China), in Australia, the world's largest lithium producer, and in other areas such as the "lithium valley" in Brazil, lithium is extracted from rock (pegmatites) mostly via open-pit mining. The frontier is also expanding across Europe. Portugal, Serbia, Germany, and Spain are thus far the main sites, but the extractive gaze is looking everywhere.[6] Extractivism is constantly inventing "lithium triangles" and similar pseudo-geographies to pitch to the financial markets to justify the endless exhumation of the earth's riches.

Environmental Destruction

Despite the term "transition" implying change, extraction is today as it always was, a process of wealth transfer from nonhuman to human communities, from poor to rich communities, from the Global South to the Global North. Resource extraction often takes place on sites of previous appropriation, due to the ease of legal and regulatory frameworks and willing governments, and in territories of poor and marginalized communities.

In the Jequitinhonha River Valley (advertised as the *lithium valley*) in Minas Gerais, Brazil, Sigma Lithium Resources—a mining company that claims to sustainably produce "green lithium"—has landed on the territories of deeply marginalized Indigenous and *quilombola*, or Afro-Brazilian, communities. It is literally on top of the Piauí River that is home to many communities that use it for recreation, fishing, foraging along its shores, and that is important spiritually. What is the meaning of the word "sustainability" here and in other such sites of extraction? None of the global communities affected by resource extraction for the green transition—be it Indigenous or traditional and agricultural communities—are expected to benefit from the commercial expansion of EVs. They "benefit" only from the destruction of their lives and futures.

It was for this reason that in 2017 I initiated an investigation into the impacts of lithium extraction on Indigenous communities and more-than-human ecologies of the Atacama Desert, the Lithium Triangle Studio.[7] At that point, there was still minimal research published on the negative impacts of lithium mining. Through multiyear remote-sensing analysis developed in collaboration with local Indigenous communities, lawyers, and archaeologists, we were able to confirm claims by local Indigenous peoples of a decrease in the water table, impacting the freshwater lagoons that surround the *salar*, due to the extraction of water for both lithium and copper mining; we confirmed how tree cover had decreased, in particular algarrobos, tamarugos, and chañar, and how the same was happening to smaller vegetation that is home to unique animal and insect life and on which donkeys and llamas feed. We observed how rare microbial ecosystems had

The proposed main pit of the Barroso mine by Savannah Resources is 500 meters away from the village of Romainho. Film still from *Inverted Mountain*, 2022.

Rock waste pile part of SIGMA lithium mine, next to houses in Quilombo Poço Dantas, Minas Gerais, Brazil, 2024.

been affected by shifts in the flow of water and by changes in the water's chemical composition, in particular the extremophile microbial mats that surround the edges of lagoons. We observed how dust and particulate materials released by mining activities generate a white haze over hundreds of kilometers, that is, permanently over the *salar*, which deposits over agricultural areas. We confirmed how water is extracted not just in the *salar*, but also upstream, near the small oases that surround it—Peine, Socaire, Camar, Toconao, and the *ayllus* of San Pedro de Atacama, small settlements that are dependent on the little water that trickles down from the top of the mountains for their agricultural survival; we noted how ground temperature had increased all across the *salar*, making agriculture much harder for the local Indigenous communities. We confirmed that mining companies hold more water rights than the *salar*'s yearly rates of replenishment, and that Atacameño or Likan-Antai communities that circumscribe the Salar de Atacama hold an amount of water rights barely sufficient for their survival.[8]

These cumulative impacts are in areas that have suffered preexisting violence, areas where some of the largest copper mining industries in the world have benefitted for decades from the "favorable business conditions" created by Augusto Pinochet and Chicago-school inspired deregulatory policies; a territory where copper mines continue expanding; a territory whose soils still hold the bodies left behind by the genocide perpetrated by Pinochet's military junta. One infamous example of this was the Water Code of 1981 that was introduced to promote private investment in mining. In addition to treating water as any other commodity, it separated it as property from the ownership of land, allowing the free buying and selling of water, without demanding compensation for the generation of adverse effects in surrounding territories. This allowed copper companies to appropriate massive amounts of water and gain control over vast areas of the desert, effectively dispossessing existing Indigenous communities.[9]

Every other mining site across the Global South tells a similar story. And those in the north are not too different. In 2022, with the Territorial Investigation Group (GIT) we analyzed the documentation submitted by Savannah Resources as part of their Environmental Impact Assessment for the Barroso Mine in Portugal. We identified impacts on vegetation and biodiversity due to the necessary stripping of soils and the need to cut down forests and scrubland to prepare for open-cast mining; the proposed destruction of precious wetlands *(lameiros)* that are important carbon sinks and sites of rich insect and vegetal life; the appropriation of hundreds of hectares of common lands, disregarding the economic livelihood they provide for the population; the construction of various water diversion and retention basins implying the destruction of aquatic ecosystems, impacting surrounding soils and vegetation; the retention of huge quantities of water when it should be flowing to surrounding soils and rivers; the high risk of contaminating groundwater, watercourses, or adjacent rivers via acid mine drainage; and a very high risk of dam collapse, with potential impacts over environments and human lives. All of this in a region that the FAO (Food and Agriculture Organization of the United Nations) recognized as World Agricultural Heritage, due to the symbiotic relationship between peoples, customs, and environment of its agro-silvo-pastoral system.

Polluted Mental Ecologies

While discourse surrounding the impacts of mining has historically centered on material aspects (e.g., the contamination of airs and rivers, dam collapses, deforestation), it is essential to consider psychological and mental dimensions as well. Long before the mines come into operation, their impact is felt on the local inhabitants, with land grabs and the environmental destruction planned for mining sites wiping out plans for other modes of development. Then there are the mining representatives who go from door-to-door coercing people to sell their lands, targeting the elderly in particular. I've witnessed this in the Atacama, in Barroso, and the same is happening now in the Jequitinhonha River Valley in Brazil. It is also common to see offices and marketing agencies being set up in the local villages as markers of mining presence. All this is accompanied by the onslaught of "expert" and opinion articles in local and national newspapers and on TV, always claiming that mining will bring jobs and take local communities out of poverty. Slowly, neighboring lands start being occupied, often illegally, and forcing communities to be constantly monitoring their surroundings. Extractive pressure comes from all sides.

After this, one can imagine the impacts on mental health of daily explosions for those who live in the vicinity of mining operations. When doing research in the Jequitinhonha, I visited the community of Poço Dantas, who have been forced to live 150 meters away from the rock waste mounds of Sigma Resources; in Portugal, the main pit of the Barroso mine by Savannah Resources is 500 meters away from the village of Romainho, and is planned to be bigger than the village itself. In both cases walls of people's houses have already suffered cracks due to the explosions.

In the Salar de Atacama the lithium extraction facilities are farther way from villages. But the impacts of extractive industries on mental ecologies is equally visible in the prevailing suspicion and intra-community conflicts. There are many reasons for this: sometimes it's about different positions on the purchase of land by mining companies; sometimes it's about the direct impact of extraction on agricultural livelihoods; sometimes it's about the heavy burden of fighting extractivism over years and decades; and sometimes it's about deciding whether to accept compensation.

Compensation (whether in the form of royalties, the construction of a school or sports field, etc.) is a common mechanism by which mining companies break down community resistance. It is seen by some as the lesser of two evils and an opportunity to benefit in at least some way from a dire situation. Mining tends to take place in the territories of marginalized communities, in contexts where centuries of dispossession and racial discrimination mean that people have very low expectations of the government or of their own ability to influence the future of their land. And so many accept compensation—often under pressure, afraid of what the future will hold. But for others, entering into negotiations with mining companies or allowing compensation is a betrayal of the struggle to protect ancestral territories, to protect life, livelihoods, and their future. All this is exacerbated by the political, legal, and financial pressure exerted by both the state and the mining companies on local leaders and representatives. Communities often collapse under the pressure of mining and its endless cash flow.

This story is common to Chile, Argentina, Portugal, Serbia, Australia, and every other place where the digging machines of extractivism have landed. In every area of resource extraction, the arrival of mining signals a drastic reduction in the range of possible futures, a trauma that affects both the environment and the people in ways that are not always visible.

Against the *Cut of Relation*

At the heart of exhuming the mineral riches of the Earth is an extractive gaze that unsees life to focus on the riches below. This unseeing is active. Impacted communities are depicted in ahistorical, static, and normative ways, their modes of existence ignored, their opacity denied. This is part of what I call the *cut of relation*, the violent attempt at separating peoples from the possibility of becoming. I speak of *relation* here in the sense given to it by Édouard Glissant, of a transformation that results

Demonstration against lithium mining in Covas do Barroso, August 2024.

from the encounter with the "other." But if *relation* speaks of a poetics of encounter, the *cut of relation* speaks of the violence of disconnection and forced isolation. The *cut of relation* is enacted as much by expert reports and environmental impact assessments that essentialize communities, as it is by media campaigns, the forced separation from land, enclosures and expulsions, or the destruction of the material and mental ecologies that are the base conditions for living. The *cut* operates at epistemological and spiritual levels with the destruction of memories, the destruction of knowledges, the disregard for imaginations. At its core, the *cut* is existential.

In other words, each new advance of the extractive frontier violently targets those whose modes of living still evade the logics of transparency that underpins the extractive gaze. In the case of the "green transition," it is targeting the modes of existence of Indigenous, agricultural, and traditional peoples across the world. Of communities that care for and sustain their environment as much as they care for and sustain themselves. Communities that have done this in the absence of government support. Communities that do not feel the need to own an EV to offset any ecological guilt complexes. Communities that in going against the logic of the state—as they always have—reaffirm the possibility of non-capitalist futures, of living in coexistence with others, the present, one's ancestors, and future generations, whether they are our own or those of other species.

That is why communities like those of the Jequitinhonha, of Barroso, of the Atacama, are resisting, building alliances with other communities across these regions and across the world, engaging in alliance-building, trans-local and rhizomatic, both rooted and in constant movement.[10] They are resisting not because they refuse transformation, but precisely the opposite, because they don't want their futures to be contained or drastically reduced by the enforcers of extraction. Resistance is nothing other than the affirmation of existence. And that is what the extractive gaze fears most.

1   Energy production corresponds to about 72 percent of global emissions. Within these, 16.2 percent refer to transportation and 12 percent correspond specifically to road transportation. Individual vehicles amount to 60 percent of road transportation emissions. Climate Watch, the World Resources Institute, 2020.

2   OPEC, "World Oil Outlook 2050," 2024, https://publications.opec.org/woo.

3   Exxon Mobil, "Global Outlook: Our View to 2050," 2024, https://corporate.exxonmobil.com.

4   The US Energy Information Administration provides a futher explanation: "By 2050, energy-related CO_2 emissions vary between a 2% decrease and a 34% increase compared with 2022 in all cases we modeled. Growing populations and incomes increase fossil fuel consumption and emissions, particularly in the industrial and electric power sectors. These trends offset emissions reductions from improved energy efficiency, lower carbon intensity of fuel mix, and growth in non-fossil fuel energy." https://www.eia.gov/todayinenergy/detail.php?id=61024.

5   Initially containing 200 to over 1,000 parts per million (ppm), the lithium brine solution is concentrated by solar evaporation to achieve a ratio of up to 6,000 ppm lithium after twelve to sixteen months. See Godofredo Pereira, "The Desert Is Not a Triangle," *Aerocene Newspaper II,* 2023, https://aerocene.org/newspaper2023-eng/.

6   Hundreds of lithium reserves have been identified for mineral prospection around Europe. Several are already undergoing EIA approval and licensing stages. Many have been highly contested by local populations and environmentalist groups, such as the Barroso and Romano mines in Portugal, the Doade and Las Navas mines in Spain, or the Jadar Project in Serbia, which sits outside of the EU but is crucial to its long-term planning. As part of the EU Critical Raw Materials Act, in March 2025, the European Commission has named twenty-five strategic extraction projects. See https://ec.europa.eu/commission/presscorner/detail/en/ip_25_864.

7   The studio was part of the MA in Environmental Architecture at the Royal College of Art, London.

8   The Lithium Triangle Research Studio, www.ea-lithiumtriangle.org.

9   See Godofredo Pereira, "Geoforensics: Underground Conflicts in the Atacama Desert," in *Forensis: The Architecture of Public Truth* (Berlin: Sternberg Press, 2014).

10 Referring to a term often used by his good friend Felix Guattari, Gissant notes how the *poetics of relation is rhizomatic,* i.e., it is rooted, not in the sense of a vertical filiation, but in the sense of a transversal spreading.

Spatial

Marina Otero Verzier in Conversation with Cara Hähl-Pfeifer, Damjan Kokalevski, Andres Lepik, and Māra Starka

Spatial

Part two of the conversation with Marina Otero Verzier discusses the location, architecture, and energy use of data centers. She argues for diverse architectural solutions that react to different types of data, depending on the local climate and spatial context. The rising demand for low-latency edge data centers and AI infrastructure has given rise to the concept of "energy geographies"—regions with abundant renewable energy, land, and favorable regulatory frameworks. The discussion calls for architects and urban planners to play a crucial role in reimagining data center typologies, recognizing the national Data Center Plan of Chile as a model that other countries can learn from.

Cara Hähl-Pfeifer, Damjan Kokalevski, Andres Lepik, and Māra Starka: As we have learned, many data centers have coded names. On the outskirts of Munich, we visited a data center by Equinix named MU4. Similarly, their centers in Amsterdam are named AM1, AM2, and so on. Why are the locations of these buildings difficult to access and understand? Is it strategically important to build them on the periphery of cities or in industrial parks?

Marina Otero Verzier: Initially, many data centers were strategically sited close to cable landing points and in regions with stable infrastructures and governments, where access to key resources—energy, water, connectivity—was relatively secure. Climate also played a role: cooler environments reduce the energy required for thermal management. These conditions contributed to the concentration of data centers in the Northern Hemisphere, particularly in parts of Europe, North America, and East Asia—regions that tend to combine temperate climates, economic strength, and advanced digital infrastructures. The result was a marked geographic imbalance, reinforcing a north-south disparity in the global digital landscape, with similar asymmetries visible within national borders. For instance, most of the data centers serving African countries were located in Marseille. This pattern, however, is evolving. The increasing demand for low-latency services—such as streaming or cloud gaming—has driven the growth of edge data centers located closer to users, including in areas not directly connected to major submarine cable routes. This has helped expand distribution networks and foster new regional clusters, where connectivity, energy availability, and financial ecosystems intersect.

At the same time, other types of data processing—especially the training of large language models (LLMs)—are less dependent on latency and can be performed in more remote or less digitally connected locations. This decoupling from cable geography, combined with the soaring energy demands of AI infrastructure, has redirected attention toward what could be called "energy geographies." Regions with abundant renewable energy, available land, and favorable regulatory frameworks—such as parts of Scandinavia, North America, and South America—are emerging as significant new nodes in the global data infrastructure. But these decisions are far from neutral. They are shaped by government incentives, land speculation, and the lobbying power of major tech companies. In many instances, infrastructure is not simply following demand but actively reshaping territorial and environmental priorities. Energy grids, water access, and public subsidies are being reorganized to support these facilities. The escalating race to build AI infrastructure has only sharpened these dynamics, raising urgent questions about ecological sustainability, public accountability, and the distribution of benefits and burdens.

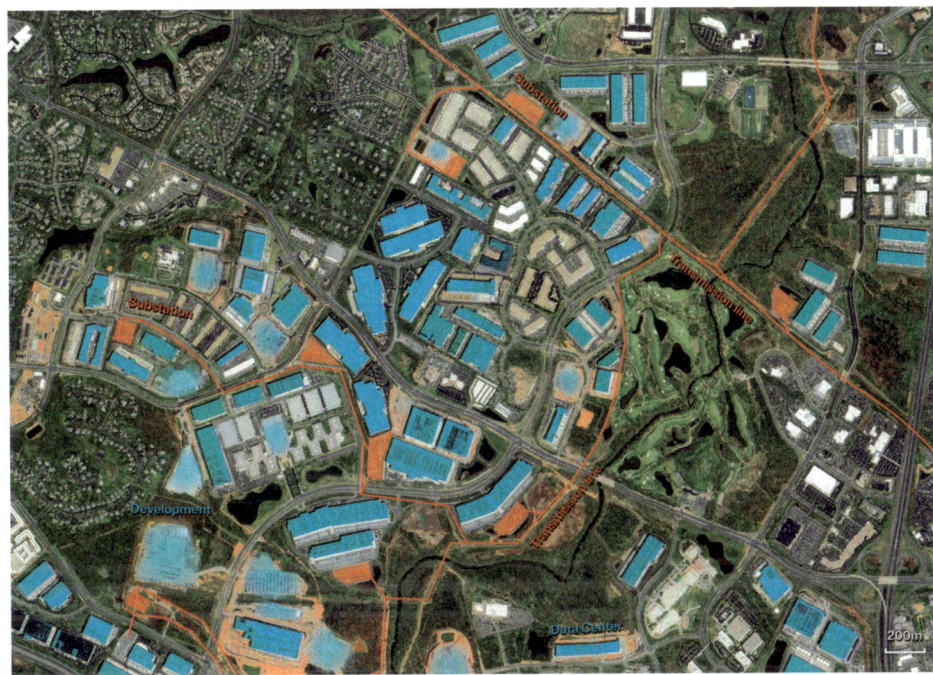

Satellite image of the Data Center Alley in Ashburn, Virginia. Data centers and development areas (blue) and power supply and distribution (orange) make up the world's densest data center cluster.

> There are estimates that if we want to achieve an AI transition, we need to build almost double the amount of data centers we have now on the planet. We currently have around 10,000 and building double that at the moment is unimaginable for energy reasons mostly. There is some rhetoric that reminds one of a new Cold War language. Not an arms race, but an AI race. This could potentially have devastating effects, but it's rhetoric we need to confront. What do you think are the mechanisms we can put in place that would require governments to think responsibly when deciding on building this infrastructure?

Too often, data center permits have bypassed standard procedures and been treated as exceptions, approved on a case-by-case basis to speed up the arrival of companies like Meta, Google, and Amazon. Just when there was mounting pressure for greater accountability and the implementation of environmental assessments, the AI race brought new urgency to attract enormous investments—among the largest in the world—and to expand infrastructure rapidly, often at the expense of proper planning.

But given the scale of operations and the volume of resources required, governments can no longer afford to make decisions one project at a time. They understand that without forward planning, they'll face serious issues with water use, energy demand, CO_2 emissions, and other environmental impacts. Some countries have already seen public protests that, in certain cases, successfully halted data center construction, including projects led by Google. This opens up important possibilities for architects and urban planners. The question is how to design data centers more responsibly, where they should be located, and what criteria should guide our decisions to ensure they do not become a threat to local communities or ecosystems. I'm seeing more openness to involving design disciplines in these conversations. Different countries are taking different approaches. Some, like the Netherlands, are focusing on clustering—creating "data center parks" around cities like

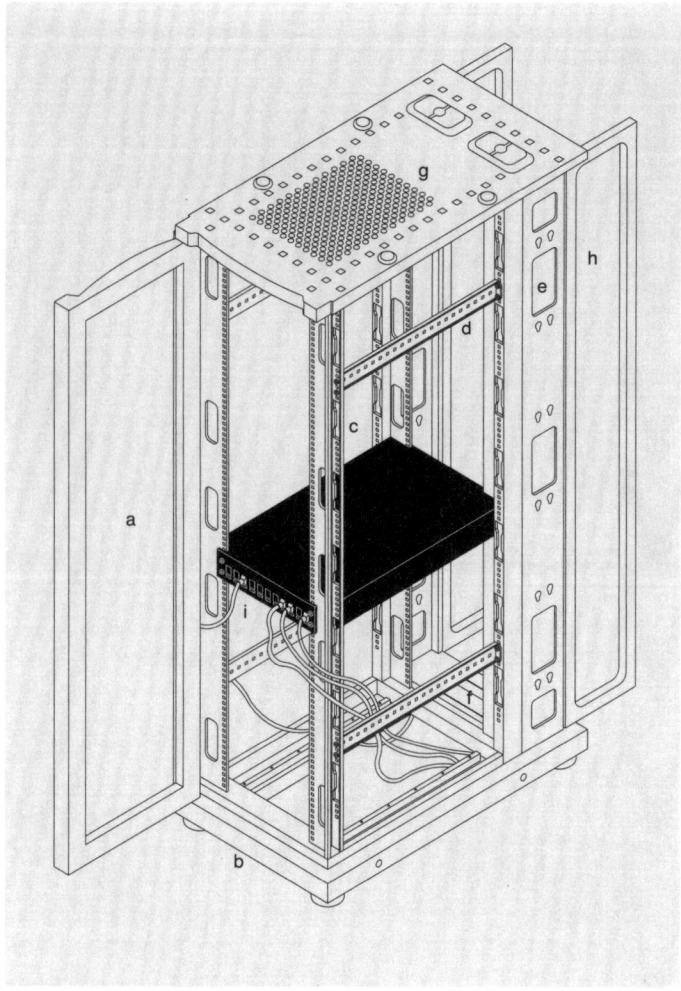

Axonometric drawing of a server rack by Maria Heinrich, 2022. It is the smallest operational unit within a data center.

a Quick-release, ventilated front door
b Front and rear skirt
c Adjustable vertical mounting rails
d Horizontal braces
e Access on side
f Access on base
g Ventilated roof
h Ventilated split rear door
i Server

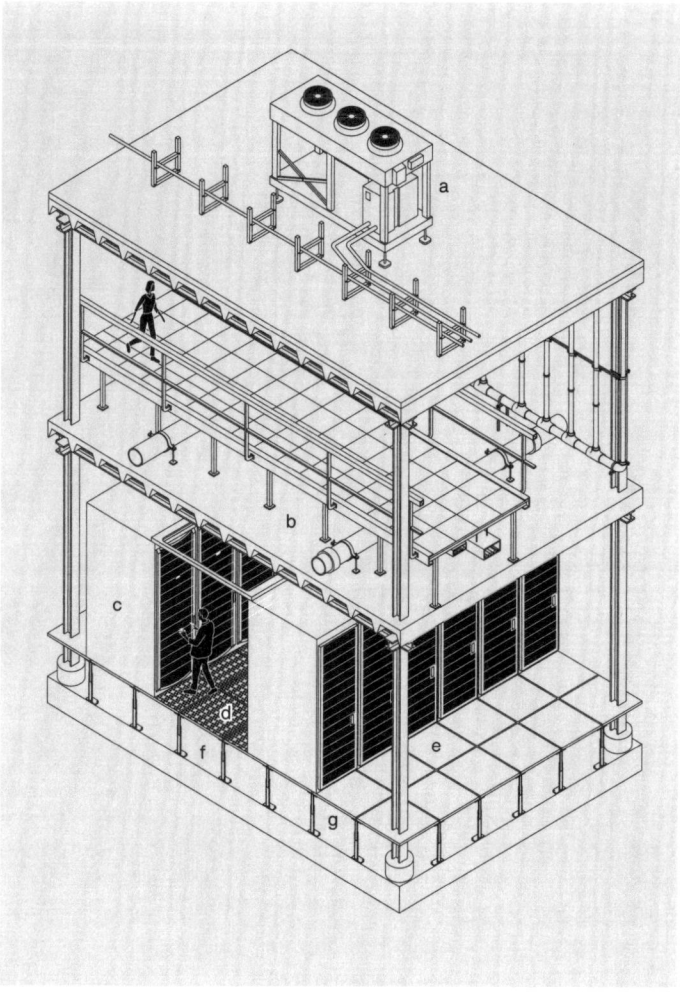

Axonometric drawing of the AMS8 Data Center in Schiphol by Maria Heinrich, 2022.

a Ventilation and cooling
b Gray space
c Server rack unit
d Hot aisle
e Cold aisle
f Perforated floor
g Raised floor

Amsterdam. This strategy allows the government to allocate power and resources more efficiently, and to better manage growth. In Spain, I have given advice to the Madrid regional government as they develop a master plan for the coming years. Surprisingly, even though the administration shows little concern for ecological consequences, it recognizes the importance of long-term planning to avoid future risks. More recently, I collaborated with the Chilean government to help design a national strategy for data centers. Everywhere, the challenge is the same: how to foster growth in the digital sector without exhausting local resources or placing the burden on nearby populations. That's exactly where architecture and planning can play a crucial role.

During the last couple of days, we visited two types of data centers in Munich: the Leibniz Supercomputing Centre (LRZ), which is publicly funded and easy to access. Then, we visited a private data center by Equinix, one of the largest interconnectivity and data center companies worldwide. Compared to the constantly adapting architecture and technological testing of the LRZ, the private data center features a clear corporate aesthetic. It's built to be invisible, with a plain, purely functional facade and barely any windows. Do you think data centers should become more visible to us as a society? Or should they be more present within the urban space?

I think it depends; the regimes of visibility vary. The quest for transparency, understood as access to and honesty in the sharing of information, is often flattened when translated directly into architectural features, like a window. But I don't necessarily believe that making something more visible makes it more democratic. Even if a data center had a "transparent" facade that let you see inside—the equipment, the servers—you still wouldn't have any agency over how those systems function. That's a trend I've been noticing more and more, for instance, in places like the Barcelona Supercomputing Center or the new IBM quantum computer in the Basque Country. Their internal systems are on view to the public, and that visibility is meant to symbolize openness or democratic values. It's a step, yes, but it doesn't resolve deeper issues: what flows through those circuits, how data is collected, managed, owned, or used.

So, I'm less concerned with how these buildings present themselves to the world, and more with how they actually operate. What matters is understanding that there are different *ecologies of data*. There's private data and public data. There's *hot* data, which needs to be processed immediately, and *cold* data, like archives, which are stored but rarely accessed. Some data is temporary, like information processed by a supercomputer that lasts ten years, or even just a few seconds, like a TikTok video. And then there's the data we, as a society, decide must be preserved for generations. If we understand those ecologies clearly, I believe they can lead to entirely different architectural approaches to data centers, ones that are more nuanced, responsive, and socially and ecologically meaningful.

> You mention the MareNostrum 4, part of the Barcelona Supercomputing Center. It has been placed in a transparent glass structure built inside a former chapel. The building is accessible to visitors who can observe the server racks and blinking lights, but it is difficult to understand its technology or what information is processed there. That's interesting in relation to our guided tour at Equinix, where we learned that they don't know what's stored on their servers, claiming it's not part of their business model. They provide storage space and sufficient security for data processing.

As we have mention earlier, there's a clear distinction between seeing a system and understanding it. That's why, when talking about data centers, I advocate for making these infrastructures more public and accountable, because they store and process collective knowledge. Still, we shouldn't aim for a single typology of data center but rather consider what kinds of typologies serve the *ecologies of data* they host. Different forms of data call for different architectural responses. If a center stores private or sensitive data, it makes sense to locate it in a secure site. But if the data is temporary, open-source, or intended for public use, perhaps the infrastructure should resemble a public

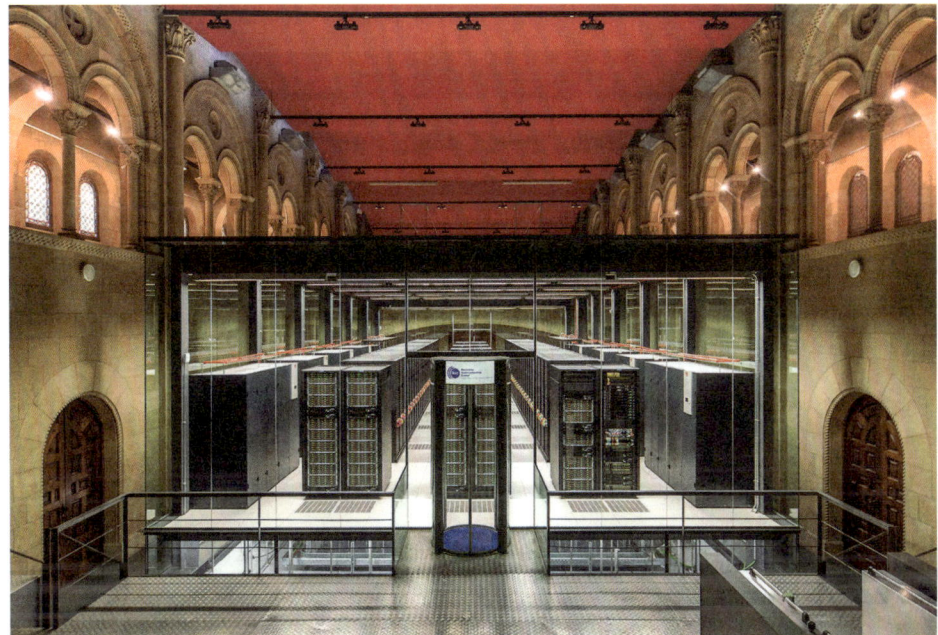

MareNostrum 4 Supercomputer inside a former chapel in Barcelona.

library. It all depends on how we relate to the data and the value we assign to it. I've also been thinking about the difference between public infrastructure and private enterprise. I believe research institutions—like LRZ—present a major opportunity to model alternative approaches. Their motivations are not strictly profit driven; they are financed with taxpayer money and serve the production of collective knowledge. That makes them ideal allies in rethinking the future of data centers, making them more transparent, more accountable, and more integrated into their social and ecological contexts.

> The private companies are focused primarily on functionality and economic gains and less so on design. However, when we are talking about buildings that are in the public realm, we should insist on a certain architectural quality and responsible design.

I totally agree with you. I think there has been some advancement in that regard. When I first started looking into data centers, we had a conversation with the Data Center Association in the Netherlands and asked: What's the role of architects in this field? Their answer was that designing the facade falls within our realm of work and expertise. Now, of course, facades matter, but design could play a much bigger role. That includes reimagining the very typology of data centers, developing planning strategies that explore hybrid uses, identifying the most appropriate areas for their integration, and designing the buildings with materials that reduce emissions. Equinix, for example, has a few well-designed "flagship" data centers. During a tour, their guide mentioned that one of their facilities was heating the swimming pool for the 2024 Paris Olympics, but he framed it as an exception. That's disappointing. Why isn't that the norm? Why aren't data centers always connected to public swimming pools or other community infrastructure? It seems like a missed opportunity. There's a lot of room for rethinking and redesigning these infrastructures from an architectural perspective.

> What kind of labor goes into maintaining these data centers? There is a narrative that they will create new job opportunities in the future; however, as we experienced during our recent visit, it was actually the opposite. There is a permanent team of around fifteen people, which is quite small. Additionally, there are external maintenance crews that are essential to the overall upkeep of the different computing and energy systems, and cleaning.

The narrative around building data centers in a region is often tied to promises of job creation. But the figures shared with the public usually include the workers needed for construction, which is a one-time job. Once the facility is operational, things look different.

Spatial

As we saw at Equinix, running a medium-scale data center requires no more than fifteen people. I've visited many data centers where only eight staff members are employed full-time. These aren't broad job creators; the positions are highly specialized. In places like Quilicura, Chile, none of the local residents have been inside the data centers, let alone been hired. There's a clear disconnect between these facilities and the surrounding communities—ironically, the same communities who bear the brunt of their presence: noise, pollution, and heavy resource use. Architecturally, data centers often reinforce this disconnect. They're designed to look closed off, highly secured, and isolated. But what this aesthetic conceals is their deep reliance on external infrastructures—water, energy, and environmental systems. They emit CO_2, generate noise, and are anything but self-contained. This dual dynamic—projecting independence while depending heavily on common resources—is deeply problematic. That's why more governments are now asking data centers to give something back: to compensate for their environmental and infrastructural impact. This might include allocating computing power to local research institutions or offering training programs for nearby communities, so residents can access future job opportunities in the sector. Still, these types of negotiations are most often incentivized by tax breaks or energy rebates offered to data center operators.

> As you pointed out, providing computing power and jobs to a community, to reciprocate the fact that their water and land resources are used, brings us back to the role of planning and architecture. Spatial planning is by definition serving a larger civic constituency, not just one company. So, maybe planning is needed to shift away from catering to the desires of large corporations for the sake of economic gains.

Just to give you an example: while working with local communities protesting against data centers in Chile, we proposed several points for inclusion in a national plan. First, data centers should not use water for cooling. Second, the energy they consume should come entirely from renewable sources. By 2035, all data centers should be energy self-sufficient—meaning they

Exterior view of the Leibniz Supercomputing Centre of the Bavarian Academy of Sciences and Humanities in Garching. The building provides space for three different functional areas: the "twin cube" in the back houses the supercomputer, as well as extensive data archives. The four-story institute building, to the front, contains the scientists' office space. The lecture hall and seminar building is located on the right. Designed by Thomas Herzog Architekten, the complex opened in 2006.

must generate their own power, ideally through solar or other renewable means. They should also be carbon-neutral. We also proposed the creation of a public information platform, where data on each data center's water and energy use, as well as emissions, would be updated monthly. Transparency is crucial, because understanding these facilities' environmental impacts is currently very difficult. In addition, environmental and social compensation strategies should be implemented in the communities hosting data centers. These strategies must create synergies—for instance, by reusing excess heat or connecting the data center's systems more productively with the local environment. We also suggested drafting a national data center law, one that guarantees community participation. That would mean some form of public consultation whenever a new data center is planned. This has been contested by the newly formed Data Center Association in Chile, which argues that such requirements would slow growth and threaten competitiveness. But our argument is simple: a region that doesn't meet these standards will never be competitive in the long run.

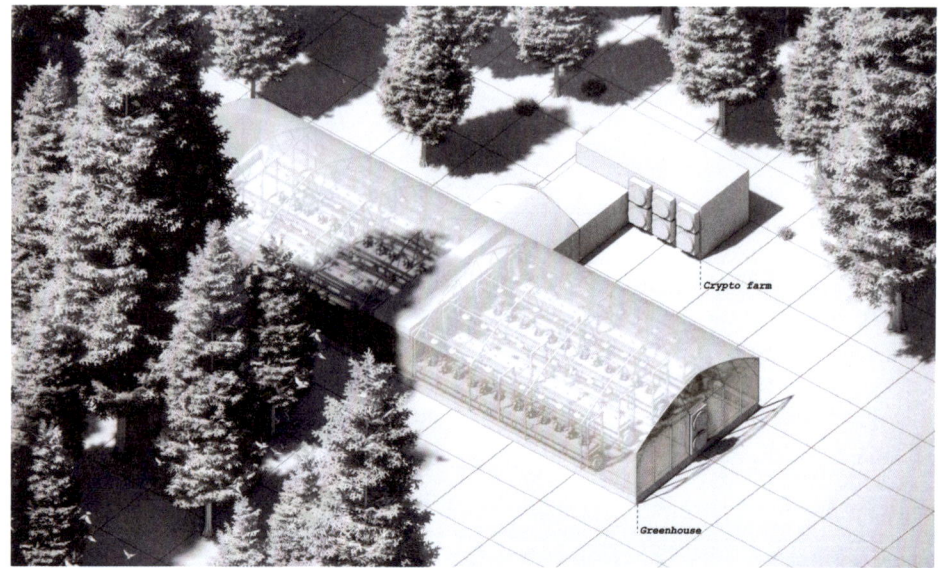

Greenhouse Project illustration by Helena Francis, 2023. The initiative by Genesis Digital Assets (GDA), in partnership with Boden Municipality in northern Sweden, is a 300-square-meter pilot greenhouse that uses waste heat from a digital currency mining data center to grow vegetables year-round.

At the same time, I've been working with architects' associations. In Madrid, for example, we're organizing training sessions to explain what data centers are, and to encourage more critical engagement with the topic. Right now, architects mostly just work from a standard blueprint and adapt it slightly—there's little room for agency or innovation. The way I managed to get their attention, a bit provocatively, was by pointing out that this is a booming sector. Millions are being invested, and architects are missing out. It was a controversial angle, but it worked—it does with architects.

> In recent years real estate agencies are discovering the development of data centers as a new site of profit making. This is based on speculation and projection, as real estate usually works. There is now another interested party introduced in this data economy.

For real estate companies, there's currently significant risk in investing in land or development for data centers, because they're unsure whether those sites—and the surrounding power grids and infrastructure—will remain viable in the near future. There's uncertainty around whether the available energy on a given plot will be sufficient to meet the growing demands of AI. In recent months, some data center developers—who saw massive profits in previous years—have found themselves out of sync with the so-called AI race. This uncertainty has reshaped the industry. Tech giants are now stepping in, proposing to build their own nuclear plants or dedicated energy systems. There's increasing talk of data center companies becoming energy providers themselves. You could quite literally call it an entanglement of power, with data centers accumulating not just information, but energy and influence.

Marija Marić

Land and Power

"Land and Power" is a "real estate poem" composed out of words and phrases sourced from real estate advertisements selling land for data center construction, published online between 2016 and 2025. It unfolds the power of language in mediating and normalizing practices of land and real estate speculation, as well as imaginaries of extraction, grounded in the settler-colonial ideology of "land improvement." In catering to the "hyper-scalers," property industries behind data center construction are themselves made in the process of hyper-scaling—of both words and footprints. Real Estate Poetry is a method of critical spatial writing and an ongoing publication project, aimed at understanding the role of property industries' language in commodification of land and housing.

virgin land [1]
unimproved land [2]
inhabited by settlers [3]
thoughtful land stewardship [4]

a premier, powered land [5]
this prime land [6]

shovel-ready industrial land [7]
rural but close to everything and everyone [8]
land for specialized developments [9]
land with endless possibilities [10]

ideal for: data centers [11]
or any operation needing expansive building footprints [12]
this property has all the elements [13]
this property is ready [14]

property comes with all sand, gravel, shale, and bluestone rights [15]
with all wind and solar rights [16]
rights to subsurface natural gas, oil, and associated hydrocarbons [17]
groundwater reserves [18]
abundant water [19]
extensive water [20]
property comes with rights [21]

a landfill decades ago [22]
accepted primarily construction and demolition debris [23]
soils that generally favor construction [24]
soils for straightforward site prep [25]
moving dirt now [26]

to the southwest, the world's largest beef processing plant [27]
to the west, reliable energy infrastructure [28]
to the east, wastewater treatment facility [29]
to the north, casino and luxury hotel [30]
located in an opportunity zone [31]

liberal zoning [32]
can be rezoned [33]
no zoning [34]
located in the enterprise zone [35]

the city has been receptive [36]
village is supportive [37]
streamlined regulatory climate [38]
situated in the extraterritorial jurisdiction [39]
you will be welcomed with open arms [40]

aggressive county incentives [41]
low real estate taxes [42]
no permits are required [43]
a rare gem [44]

great demographics [45]
above-average household income [46]
a population 55% of whom hold a bachelor's degree or higher [47]
robust talent pool ensures a steady pipeline of tenant operations [48]
affluent communities [49]
situated in the path of development [50]

currently leased [51]
has two homes on it occupied by owners and farm land [52]
a minor waterway that may need to be relocated [53]
no known restrictions [54]
shovel ready [55]
and just waiting for development [56]

a substation [57]
fiber backbone [58]
both lit and dark [59]
hyper fiber [60]
supportive fiber [61]
infrastructure in place [62]

miles of river frontage [63]
miles of railroad frontage [64]
miles of road frontage [65]
infrastructure ensures redundancy [66]

at the very heart [67]
in the middle of [68]
within a stone's throw [69]

an area of explosive growth [70]
scalable [71]
abounding [72]
the fastest growing corridor [73]
wake up to this incredible parcel [74]

land for hyperscale data center development [75]
high tension power lines [76]
for high energy users [77]
power-hungry operations [78]
high-power needs [79]

development is roaring [80]
data center development [81]

act now [82]

1. "10230 Panoramic Dr: 24.35 Acres of Industrial Land in Shiller Park, IL," LoopNet, accessed May 18, 2025, https://www.loopnet.com/Listing/10230-Panoramic-Dr-Schiller-Park-IL/35003755/.

2. "Scottsville Road – 1,296 +/- ACRE FOR DATA CENTER POWER PLANT: 1,296 Acres of Commercial Land Offered at $15,900,000 in Mehoopany, PA 18629," LoopNet, accessed March 25, 2025, https://www.loopnet.com/Listing/Scottsville-Road-Mehoopany-PA/34858300/.

3. Ibid.

4. "The Lasting Benefits of Data Centers," Tract, accessed April 16, 2025, https://www.tract.com/our-approach/.

5. "460 County Rd – Powered Land/ Potential Data Center: 100 Acres of Industrial Land in Elgin, TX 78621," LoopNet, accessed March 25, 2025, https://www.loopnet.com/Listing/460-County-Rd-Elgin-TX/34386682/.

6. "21201 Morgan Rd – GREAT DATA CENTER LOCATION: 10.63 Acres of Commercial Land Offered at $2,950,000 in Land O Lakes, FL 34639," LoopNet, accessed March 25, 2025, https://www.loopnet.com/Listing/21201-Morgan-Rd-Land-O-Lakes-FL/24419007/.

7. "11515 Derry Rd & 7314 Sixth Line, 142.48 Acres of Industrial Land Offered at $182,379,143 USD in Milton, ON L9T 7J5," LoopNet, accessed March 25, 2025, https://www.loopnet.com/Listing/11515-Derry-Rd-7314-Sixth-Line-Milton-ON/34437962/.

8. "I-95 & Ladysmith Rd: 45 Acres of Commercial Land Offered at $5,000,000 in Ladysmith, VA 22501," LoopNet, accessed March 25, 2025, https://www.loopnet.com/Listing/I-95-Ladysmith-Rd-Ladysmith-VA/21959447/.

9. "2820 Nokomis rd – 2820 Nokomis Rd: 11.27 Acres of Commercial Land Offered at $778,890 in Lancaster, TX 75146," LoopNet, accessed March 25, 2025, https://www.loopnet.com/Listing/2820-Nokomis-rd-Lancaster-TX/35025102/.

10. "100 Volunteer Way: 19.52 Acres of Commercial Land Offered at $3,000,000 in Waretown, NJ 08758," LoopNet, accessed March 25, 2025, https://www.loopnet.com/Listing/100-Volunteer-Way-Waretown-NJ/34144144/.

11. "460 County Rd – Powered Land/Potential Data Center," accessed March 25, 2025.

12. Ibid.

13. "8274 Maple Tree Ln – Maple Tree Farm | Warrenton: 145 Acres of Agricultural Land in Warrenton, VA 20187," LoopNet, accessed March 25, 2025, https://www.loopnet.com/Listing/8274-Maple-Tree-Ln-Warrenton-VA/27699702/.

14. "1751 Orbit Way: 8.77 Acres of Industrial Land Offered at $999,000 in Minden, NV 89423," LoopNet, accessed March 25, 2025, https://www.loopnet.com/Listing/1751-Orbit-Way-Minden-NV/33237800/.

15. "Scottsville Road – 1,296 +/- ACRE FOR DATA CENTER POWER PLANT," accessed March 25, 2025.

16. Ibid.

17. Ibid.

18. Ibid.

19. "8537 Braun Rd – Land for Development – Mt. Pleasant, WI: 140 Acres of Industrial Land in Mount Pleasant, WI 53403," LoopNet, accessed March 25, 2025, https://www.loopnet.com/Listing/8537-Braun-Rd-Mount-Pleasant-WI/34625759/.

20. Ibid.

21. "Scottsville Road – 1,296 +/- ACRE FOR DATA CENTER POWER PLANT," accessed March 25, 2025.

22. "10230 Panoramic Dr: 24.35 Acres of Industrial Land in Schiller Park, IL 60131," accessed March 25, 2025.

23. Ibid.

24. "460 County Rd – Powered Land/ Potential Data Center," accessed March 25, 2025

25. Ibid.

26. "Ernest M Smith Blvd – Stuart Crossing- Parcels A, E, F, and J: 9.88 – 20 Acre Commercial Land Lots Offered at $2,152,000 – $3,456,000 Per Lot in Bartow, FL 33830," LoopNet, accessed March 25, 2025, https://www.loopnet.com/Listing/Ernest-M-Smith-Blvd-Bartow-FL/30976034/.

27. "5208 Dakota Ave – Industrial Land Behind Circle S South Station: 31.74 – 40 Acre Industrial Land Lots Offered at $8,295,566 – $10,454,400 Per Lot in South Sioux City, NE 68776," LoopNet, accessed March 25, 2025, https://www.loopnet.com/Listing/5208-Dakota-Ave-South-Sioux-City-NE/25000722/.

28. Ibid.

29. Ibid.

30. Ibid.

31. "FM 1660 & FM 973 – Panorama Taylor: 2.1 – 64.13 Acre Commercial Land Lots Offered at $914,760 – $12,264,433 Per Lot in Taylor, TX 76574," LoopNet, accessed March 25, 2025, https://www.loopnet.com/Listing/FM-1660-FM-973-Taylor-TX/32045608/.

32. "130 E Sydney Dr – Sydney Dr./E of USA Pky. 5 Acres of Industrial Land in Mccarran, NV 89434," LoopNet, accessed March 25, 2025, https://www.loopnet.com/Listing/130-E-Sydney-Dr-Mccarran-NV/34370984/.

33. "42649 Hearford Ln – Goose Creek: 9.62 Acres of Residential Land Offered at $10,000,000 in Ashburn, VA 20147," LoopNet, accessed March 25, 2025, https://www.loopnet.com/Listing/42649-Hearford-Ln-Ashburn-VA/13458635/.

34. "17115 N Lund Rd – 83.9 Acres in Coupland TX! 84 Acres of Commercial Land in Coupland, TX 78615," LoopNet, accessed March 25, 2025, https://www.loopnet.com/Listing/17115-N-Lund-Rd-Coupland-TX/33994917/.

35. "1205-1297 NE Evergreen Rd – Jackson South Corporate Park: 32.12 Acres of Industrial Land in Hillsboro, OR 97124," LoopNet, accessed March 25, 2025, https://www.loopnet.com/Listing/1205-1297-NE-Evergreen-Rd-Hillsboro-OR/35125676/.

36. "Highway 317 & West Adams, Temple, TX 76502," LoopNet, accessed March 25, 2025, https://www.loopnet.com/Listing/Highway-317-West-Adams-Temple-TX/25675692/.

37. "Somers Rail & Commerce Park I: 181 Acres of Commercial Land Offered at $27,150,000 in Kenosha, WI 53144," LoopNet, accessed March 25, 2025, https://www.loopnet.com/Listing/Somers-Rail-Commerce-Park-I-Kenosha-WI/28802986/.

38. "460 County Rd – Powered Land/ Potential Data Center," accessed March 25, 2025.

39. Ibid.

40. "Lot #00900 Commerce – Two Tom McCall Business Park Lots 00900/00901: 1.67 Acre of Industrial Land Offered at $345,538 - $345,538 Per Lot in Prineville, OR 97754," LoopNet, accessed March 25, 2025, https://www.loopnet.com/Listing/Lot-00900-Commerce-Prineville-OR/28391206/.

41. "I-95 & Ladysmith Rd," accessed March 25, 2025.

42. "Somers Rail & Commerce Park I," accessed March 25, 2025.

43. "Blue Mound Rd & John Day Rd – Raw Land 1 to 20 Acres Commercial/Industrial: 97 Acres of Commercial Land in Haslet, TX 76052," LoopNet, accessed March 25, 2025, https://www.loopnet.com/Listing/Blue-Mound-Rd-John-Day-Rd-Haslet-TX/31004708/.

44. "1751 Orbit Way," accessed March 25, 2025.

45. "6345 Boynton Beach Blvd: 4.5 Acres of Commercial Land Offered at $9,375,000 in Boynton Beach, FL 33437," LoopNet, accessed March 25, 2025, https://www.loopnet.com/Listing/6345-Boynton-Beach-Blvd-Boynton-Beach-FL/25179530/.

46. Ibid.

47. "Class A Corporate Campus Broomfield, CO5 Properties, Online Auction Sale, Broomfield, CO 80021," LoopNet, accessed March 25, 2025, https://www.loopnet.com/Listing/Class-A-Corporate-Campus-Broomfield-CO/34256700/.

48. Ibid.

49. "Blue Mound Rd & John Day Rd – Raw Land 1 to 20 Acres Commercial/Industrial," accessed March 25, 2025.

50. "0 Refugee Rd: 104 Acres of Industrial Land in Pataskala, OH 43062," LoopNet, accessed March 25, 2025, https://www.loopnet.com/Listing/0-Refugee-Rd-Pataskala-OH/31713289/.

51. "19810 Janak Rd: 49.89 Acres of Residential Land Offered at $2,975,000 in Coupland, TX 78615," LoopNet, accessed March 25, 2025, https://www.loopnet.com/Listing/19810-Janak-Rd-Coupland-TX/33441642/.

52. "16675 FM 1660 – 38+/- Acres Near Samsung/Megasite: 38 Acres of Commercial Land Offered at $7,500,000 in Taylor, TX 76574," LoopNet, accessed March 25, 2025, https://www.loopnet.com/Listing/16675-FM-1660-Taylor-TX/31239939/.

53. "5010 Palmer Rd *DATA*-(Future)/PWR-GEN./INDUS./RETAIL/MIXED: 73 Acres of Commercial Land in Millersport, OH 43046," LoopNet, accessed March 25, 2025, https://www.loopnet.com/Listing/5010-Palmer-Rd-Millersport-OH/33598360/.

54. "19810 Janak Rd," accessed March 25, 2025.

55. "Lot #00900 Commerce – Two Tom McCall Business Park Lots 00900/00901," accessed March 25, 2025.

56. Ibid.

57. "2212 Alliance Rd – Old Cooper-Marine Chip Mill & Barge Loading: 70 Acres of Industrial Land Offered at $5,000,000 in Quinton, AL 35130," LoopNet, accessed March 25, 2025, https://www.loopnet.com/Listing/2212-Alliance-Rd-Quinton-AL/20969397/.

58. "455 Italy dr: 12.81 Acres of Industrial Land Offered at $4,000,000 in Sparks, NV 89437," LoopNet, accessed May 18, 2025, https://www.loopnet.com/Listing/455-Italy-dr-Sparks-NV/35322429/.

59. "Urbana Pike – Industrial Land Zoned for Data Center/Biotech90 Acres of Industrial Land in Frederick, MD 21754," LoopNet, accessed March 25, 2025, https://www.loopnet.com/Listing/Urbana-Pike-Frederick-MD/18151857/.

60. "I-95 & Ladysmith Rd," accessed March 25, 2025.

61. "Urbana Pike – Industrial Land Zoned for Data Center/Biotech90 Acres of Industrial Land in Frederick, MD 21754," accessed March 25, 2025.

62. "Corners of I-555 and Highway 149 2 Large Tracts with Interstate Presence: 70 – 120 Acre Industrial Land Lots Offered at $1,050,000 – $1,800,000 Per Lot in Marked Tree, AR 72365," LoopNet, accessed March 25, 2025, https://www.loopnet.com/Listing/Corners-of-I-555-and-Highway-149-Marked-Tree-AR/8487690/.

63. "Scottsville Road – 1,296 +/- ACRE FOR DATA CENTER POWER PLANT," accessed March 25, 2025.

64. Ibid.

65. Ibid.

66. "460 County Rd – Powered Land/Potential Data Center," accessed March 25, 2025.

67. "4825 NE Starr Blvd – 3.34 Acre Hillsboro, Oregon Commercial Site: 3.34 Acres of Commercial Land Offered at $1,950,000 in Hillsboro, OR 97124," LoopNet, accessed March 25, 2025, https://www.loopnet.com/Listing/4825-NE-Starr-Blvd-Hillsboro-OR/31280364/.

68. "E 11th St: 38.89 Acres of Commercial Land Offered at $49,500,000 in Catoosa, OK 74015," LoopNet, accessed March 25, 2025, https://www.loopnet.com/Listing/E-11th-St-Catoosa-OK/31643803/.

69. "Lot #00900 Commerce – Two Tom McCall Business Park Lots 00900/00901," accessed March 25, 2025.

70. "25605 W 111th St – 114 acre research/light industrial site: 114 Acres of Agricultural Land in Plainfield, IL 60585," LoopNet, accessed March 25, 2025, https://www.loopnet.com/Listing/25605-W-111th-St-Plainfield-IL/30044062/.

71. "5208 Dakota Ave – Industrial Land Behind Circle S South Station," accessed March 25, 2025.

72. "Blue Mound Rd & John Day Rd – Raw Land 1 to 20 Acres Commercial/Industrial," accessed March 25, 2025.

73. "State Highway 151 & Military Drive W – ±19.29 acres | SH 151, Military Dr, Escala Pk 1.18 – 2.88 Acre Commercial Land Lots in San Antonio, TX 78251," LoopNet, accessed May 18, 2025, https://www.loopnet.com/Listing/State-Highway-151-Military-Drive-W-San-Antonio-TX/35161002/.

74. "4825 NE Starr Blvd – 3.34 Acre Hillsboro, Oregon Commercial Site: 3.34 Acres of Commercial Land Offered at $1,950,000 in Hillsboro, OR 97124," LoopNet, accessed March 25, 2025, https://www.loopnet.com/Listing/4825-NE-Starr-Blvd-Hillsboro-OR/31280364/.

75. "5208 Dakota Ave – Industrial Land Behind Circle S South Station," accessed March 25, 2025.

76. "Somers Rail & Commerce Park I," accessed March 25, 2025.

77. Ibid.

78. "460 County Rd – Powered Land/ Potential Data Center," accessed March 25, 2025.

79. "795 Windsweep Farm Rd – Data Center Site: 1,443 Acres of Industrial Land in Thomaston, GA 30286," LoopNet, accessed March 25, 2025, https://www.loopnet.com/Listing/795-Windsweep-Farm-Rd-Thomaston-GA/33786730/.

80. "Blue Mound Rd & John Day Rd – Raw Land 1 to 20 Acres Commercial/Industrial," accessed March 25, 2025.

81. "5208 Dakota Ave – Industrial Land Behind Circle S South Station," accessed March 25, 2025.

82. "E Jefferson Avenue – Laporte 55 Acre Residential Development Land: 55 Acres of Residential Land Offered at $2,450,000 in La Porte, IN 46350," LoopNet, accessed March 25, 2025, https://www.loopnet.com/Listing/E-Jefferson-Avenue-La-Porte-IN/32149348/.

Niklas Maak

Where the Cloud Becomes Reality:
On the Iconography of the Data Center

Aerial view of the Colossus data center by xAI in southwest Memphis, TN, United States.

Data centers are the largest and most expensive buildings of our time. Their architecture reveals a lot about the state of our society.

The Biggest Buildings of Our Time

Few building typologies have experienced a comparable boom in recent years as the data center—often called a "server farm," evoking an image of computer server racks lined up in rows like digital-age cattle in a stable. According to studies, the market for data center construction will be worth up to $79 billion[1] in the coming years; in the United States alone, over 5,000 large-scale buildings have been constructed to date, not including the countless smaller server facilities.[2] The building boom has a clear driver: every text message or email sent triggers a server in a data center somewhere, springing into action. Artificial intelligence (AI) consumes as much energy as a large city, fueling the rise of entire digital landscapes—vast clusters of computers where only data resides.

"Server Farm" and "Cloud": On the Semiotics of an Ideological Form

The "cloud" is a central ideological term in the semiotics of the new data center typology. The name "cloud" suggests an ethereal lightness that contradicts the fact that the internet and artificial intelligence consume more energy than international air travel. The concept of the "cloud" carries a reassuring promise—that our data is safeguarded in a seemingly celestial way, transcending into eternal knowledge. Floating in a global, placeless space, it feels untouchable, indestructible, immune to erasure.

Customers of the OVH data center were shocked to realize that the "cloud" they had relied on was, in reality, nothing more than a vast warehouse of servers—vulnerable to destruction by a single fire. On the night of March 10, 2021, two of the four data centers operated by Europe's largest hosting provider went up in flames. Among the lost data were critical records from government organizations, including the website of the City of Colmar. In total, 3.6 million websites went offline—banks, news outlets, official portals, and even the French government's open-data site data.gouv.fr. Several OVH clients, seeking to cut costs, had opted not to synchronize their data. As a result, their digital archives burned beyond recovery, as irretrievable as paper records turned to ash.[3] The night of the Strasbourg fire made one thing painfully clear: the data storage technology often portrayed as an untouchable force of nature is, in reality, grounded in physical hardware and just as vulnerable. The pristine "cloud" had turned into a foreboding column of smoke. Yet, even without disasters like these, the cloud has a dirty side: more than a billion people search Google every day, interact with ChatGPT, upload videos, and generate 5.8 billion likes. And despite ongoing efforts to achieve climate neutrality, the sheer storage demands of artificial intelligence have driven the energy consumption of purpose-built server farms to staggering levels.[4]

Spatial

Major fire at the data center cluster OHV in Strasbourg, France, on March 10, 2021.

The Foundations of the Digital World

The largest buildings of our time are no longer skyscrapers or malls; they are data centers. Most server farms are both massive and virtually invisible, hidden in plain sight. That holds true for the facility Elon Musk built in Memphis, Tennessee, in 2024 for his startup xAI. Its ambitions are revealed only by its name—Colossus—and the sheer scale of its infrastructure: 100,000 Nvidia Hopper H100 processors. Within Colossus, the next iteration of xAI's language model, Grok 3, is set to be trained.[5]

Colossus is a symptomatic building of our time—erected by the world's richest man, emblematic of the immense profits digitalization can generate, and, at the same time, a striking example of the problematic entanglement between private corporate interests and governmental responsibilities. As the Trump administration's former deregulation chief, Elon Musk embodies this uneasy fusion of entrepreneurship and state authority.[6]

Musk's data center sparked controversy even before its opening. Activists and the Southern Environmental Law Center have criticized both the use of public funds for its construction and its staggering energy consumption—150 megawatts, equivalent to the power needs of 100,000 average households.[7] Beyond its environmental impact, Musk's Colossus raises deeper concerns about democracy and individual autonomy. His company, xAI, trains its AI model, Grok, using posts from users on X (formerly Twitter)—without their explicit consent. Rarely has an architecturally unremarkable structure so clearly embodied two of the most pressing issues of our time: the enormous energy consumption that makes AI an environmental threat and the erosion of government control as a handful of tech oligarchs consolidate power. These figures amass billions of dollars by exploiting user data and manipulating digital discourse. The name Colossus itself signals a grandiose assertion of dominance, steeped in antiquated rhetoric. It is no coincidence that Musk posed for photos[8] at the Colosseum in Rome, where he famously proposed a physical showdown with Facebook founder Mark Zuckerberg—as if the two were gladiators of the digital age.[9]

The Ecological Problem

According to Frankfurt's climate protection report, the city may fail to meet its 2050 energy targets due to the power demands of its servers. In 2020 alone, Frankfurt's data infrastructure consumed 1,600 gigawatt-hours of electricity—60 percent more than the total consumption of its 400,000 households.[10] Globally, the information and communication technology sector now accounts for at least 2 to 4 percent of greenhouse gas emissions, a figure that continues to rise—surpassing even the emissions from global air travel.[11] In the United States, large-scale data centers consume between 313 and 509 million liters of water annually, depending on the calculation method.[12] If the internet were a country, its electricity consumption and carbon emissions would rank just behind the United States and China.

Electronic Numerical Integrator and Computer in Philadelphia, Pennsylvania. Glen Beck (background) and Betty Snyder (foreground) program the ENIAC in building 328 at the Ballistic Research Laboratory (BRL) for the calculation of ballistic tables of the US Army.

The Data Center as a Symbol of National Progress

The first true data center was established in 1946 at the Moore School of Electrical Engineering in Philadelphia, Pennsylvania. There, physicist John William Mauchly and engineer John Presper Eckert unveiled a 167-square-meter room filled with machinery—its towering server racks already evoking the monumental scale of today's data centers.[13] The Electronic Numerical Integrator and Computer (ENIAC), a $400,000 project, was originally designed to assist the US military in calculating ballistic trajectories during World War II. However, it was only completed after the war ended. The massive machine weighed 27 tons and housed 17,468 vacuum tubes, many of which frequently failed, leading to calculation errors. It also contained 7,200 diodes and consumed 150 kilowatts of power—an enormous amount for its time. Yet, when operational, ENIAC could perform 5,000 calculations per second, a staggering 1,000 times faster than conventional computing methods of the era.

Even in East Germany, the state positioned itself as a driving force in digital policy. Historian Martin Schmitt[14] examines the history of the Potsdam Data Center (DVZ), showing how the GDR used striking modernist architecture for data storage and analysis to project an image of technological progress. Equipped with the Robotron 300 electronic computer, the center processed everything deemed essential for the advancements of the GDR and its citizens—from food supply calculations to official cancer statistics, from financial transactions to domestic trade, from cadastral data to housing needs. In 1981, the Potsdam Data Center even developed a database for dietary recipes. Built between 1969 and 1972 by a collective led by Sepp Weber for the VEB Maschinelles Rechnen, the data processing center stood in the heart of Potsdam's historic district. Architecturally ambitious, its facade featured vertical slats, echoing the columns and pilasters of Potsdam's classical architecture—framing the computer age as the antiquity of the new human era. The classical motif extended to the use of mosaic art, applied to depict visions of the future. The building's base was adorned with an eighteen-piece glass mosaic by the artist Fritz Eisel, illustrating applications of electronic data processing. The scenes in this visual narrative, titled *Man Conquers the Cosmos,* resemble Latin American murals, depicting stages of scientific and technological evolution—from agricultural machinery to space exploration and computing. One of the panels prominently features a Robotron 300 system. Data centers were central to the GDR's vision of the future —just as large-scale computers were to Chilean President Salvador Allende, who sought to use them to manage his country's economy starting in 1970.

British cybernetician Stafford Beer, creator of the Viable System Model and a pioneer in the study of adaptability in biological and social systems, envisioned a groundbreaking digital infrastructure for Chile. His plan was to equip 400 nationalized factories with individual computers, allowing real-time transmission of material and energy requirements via telephone lines to a central computer housed in President Salvador Allende's government palace. At the heart of this system was Cybersyn, a futuristic data center designed by German designer Gui Bonsiepe in the aesthetic of space-age modernism. The software Cyberstride was intended to coordinate factory operations, making resource distribution more efficient and predictable.[15]

The project faced major difficulties, rooted more in Chile's financial constraints than in Stafford Beer's ideas. With the 1973 military coup against Allende, Cybersyn was abruptly dismantled—along with the vision of treating data as a "common good" rather than the property of individual corporations.

Cybersyn represented a futuristic ideal, promising emancipation and prosperity for all through cybernetic, real-time digital data analysis. Despite its shortcomings, many researchers today consider it groundbreaking—not only because it was designed around the needs of the population, but also because Beer sought to organize real-time interaction between government and citizens, anticipating key concepts of the modern internet. One wall in the operations room in Santiago was dedicated to Project Cyberfolk, an initiative aimed at measuring the collective satisfaction of Chile's population. Half a century later, all that remains of this vision of interactive political engagement is the "like" button—and a blue Facebook thumbs-up emoji.

The mosaic *Der Mensch bezwingt den Kosmos* (Humanity Conquers the Cosmos) by Fritz Eisel at the base of the data processing center (DVZ) in Potsdam, GDR, 1972.

Operation room of the project Cybersyn in Chile.

Architecture of Invisibility:
The Politics of the Anonymous Box

After the collapse of Allende's utopian vision, a global shift toward the privatization of data and its analysis took hold. As internet theorist Evgeny Morozov explains,[16] members of Stafford Beer's project moved to the then-emerging Silicon Valley, where they played a key role in developing the systems of analysis, monetization, and manipulation that now define every data-driven economy.

Data centers followed a trajectory similar to the internet itself—once envisioned as a hippie-like countercultural tool for emancipation, it was soon reshaped by commercial interests and corporate strategies.

Increasingly operated by private entities, data centers ceased to be symbols of public welfare policy and state-driven innovation. Instead, they became mere storage facilities for data, designed with pragmatic efficiency rather than architectural ambition. As the forces behind data processing grew more focused on behavioral analyses and predictive modeling, the buildings of data storage and analysis became increasingly invisible.

Server farms are the sites where high-stakes data manipulation takes place. Before the 2016 US presidential election, Cambridge Analytica collected and analyzed the personal data of eighty-seven million Facebook users[17]—along with their friends—through the quiz app This Is Your Digital Life. The insights gained were used to craft psychologically tailored campaigns for Republican presidential candidates Ted Cruz and Donald Trump.

Snøhetta's design proposal "The Spark."

At the same time, the demand for data storage surged. The larger the datasets required for Big Data, cloud computing, and artificial intelligence, the more data centers had to be built. The cloud business—the outsourcing of IT computing power and services to third-party providers—has become a major driver of corporate profits. For Amazon, cloud services are a key pillar of its financial success, with analysts estimating that Amazon Web Services (AWS) accounts for the largest share of its more than $2 trillion market capitalization. Microsoft's Azure ranks as the second-largest hyper-scaler, followed by Google Cloud in third place.[18]

The deep entanglement between the military and private tech corporations has become increasingly problematic. AWS has signed a $10 billion contract with the Pentagon for the Joint Warfighting Cloud Capability (JWCC).[19] Microsoft, Google, and Oracle are also involved, making the military dependent on these companies for its operations. This reliance raises concerns that individual entrepreneurs—such as Elon Musk, Jeff Bezos, and Palantir CEO Alex Karp—could potentially disrupt an entire nation's defense infrastructure. Social media giants not only shape political opinions and decisions but also encroach on sovereign functions, gradually hollowing out the state. At the same time, they collect user data, turning their data centers into hubs for prediction, manipulation, and control—the places where our digital avatars reside.

As data centers become increasingly invisible, and as political and economic power structures dissolve into the rhetoric of the cloud, critical questions fade from public discourse: Where do the trillions of dollars generated from selling and analyzing user data actually go? Who will set the future rules for data processing? Could a model of public, collective data ownership—one that denies access to corporations and governments—serve as the foundation for a more equitable, less growth-driven society, where sovereign responsibilities remain in public hands rather than being outsourced to private entities?

Iconography of Disappearance:
Camouflaging the Data Center as Nature

One of the iconographic strategies used by tech corporations to soften the image of data centers is their integration into nature. The architectural firm Snøhetta presents the server farm as a holistic outdoor experience, avoiding concrete and steel in favor of locally sourced stone for its structural framework. With its "Spark" project, Snøhetta aims to redefine the role of data centers as "anchors for smart city developments" while also "restoring the human touch to a digital world dominated by smartphones." According to the architects, the goal is to recenter the human body within a "living, breathing city."[20] This marks a new stage in the narrative surrounding server farms—intended to repair society and the bodies distorted by digital life. In this vision, the server farm becomes both the "body and brain" of the new city.

Iconography of Power:
The New Visibility of Digital Space

We are witnessing a new era of visibility. Tech giants and data center operators have abandoned their former discretion. Much like medieval castles or the grand headquarters of major banks, server farms are now being showcased as the treasure vaults of the digital age—the places where the most valuable assets of our time are created.

Qianhai Telecommunication Center, Shenzhen, China.

Images of Progress

A new wave of futuristic optimism is shaping the design of data centers, framing digital technology as an opportunity for progress. One striking example is the Qianhai Telecommunication Center, a 110-meter-high tower designed by Schneider + Schumacher for Shenzhen, China.[21] With 3,515 high-performance servers, it is a monument to data storage. The building is largely windowless, and its facade is adorned with thin steel elements, arranged to represent the number pi in binary code. The zeros move with the wind, creating the illusion of a living surface—an architectural spectacle that transforms the data processing into a visual experience.

The growing involvement of architects in shaping data centers as symbols of power in the digital age is reflected in numerous design awards. In the United States, Quality Uptime Services sponsors the Data Center Architectural Award, while the online magazine *Data Center Dynamics* honors multiple projects. Their website states: "Data centers should look good," showcasing the "Top 10 beautiful data centers." Among them is the Switch Pyramid near Grand Rapids, Michigan, a 21,000-square-meter data center shaped like a seven-story pyramid. Originally designed by the office furnishing company Steelcase as a design hub, its repurposing lends the facility pharaonic grandeur.[22] The data stored within is now presented as a treasure trove, echoing the sacred significance of Egyptian tombs.

For a Politicization of the Data Center

If data is the greatest collective asset of a digital society—the gold, oil, and raw material of the twenty-first century, the foundation of both economy and politics—and if even government functionality depends on access to this data, which is increasingly harvested by private platforms, then shouldn't at least some data be treated as commons, part of the public infrastructure? Years ago, internet pioneers Evgeny Morozov and Francesca Bria argued that "city governments must retain ownership of their own data and the information generated by residents, rather than handing it over carelessly to private entities." They emphasized that municipalities should "control critical infrastructures (software, hardware, datacentres), and collaborate to reduce dependence on major tech corporations in artificial intelligence and machine learning. Such steps ought to allow them to follow a pathway towards technological sovereignty." According to Bria, we would need "data commons—collective datasets—that we can use to create public value," warning that without such an approach, governments would lose their expertise and their ability "to manage and shape society in their own interest—rather than in the interest of the tech companies."[23]

Francesca Bria gained international recognition in 2015 when, as a member of Barcelona's city government, she launched the world's largest experiment in digital democracy. That year, 400,000 citizens voted on municipal online platforms to shape policies of housing and transportation. Nowhere else had citizen preferences been translated into policy so swiftly, radically transforming both the city's landscape and the concept of urban planning itself. This experiment was made possible by the DECODE Project, an EU-funded initiative Bria had previously launched in London. DECODE aimed to restore citizens' control over their data, challenging conventional notions of governance and civic participation. European developers working on the project successfully created algorithms that allowed people to decide which of their data to share, with whom, and which to keep private. "The EU could say: The data generated in Europe by our citizens is a public good—it cannot be stolen. If you want to use some of it, you have to pay us," Bria argued. "Right now it's the other way around: We give away our data for free, and then we also pay for the services that tech companies distill from it. So, in effect, we are paying twice."[24]

In Barcelona, Bria successfully pushed for a contractual agreement with Vodafone, the city's telecommunications provider, ensuring that publicly generated data would not be exploited for corporate purposes but instead anonymized and published on the municipal open-data portal. "This was about bicycle lanes, spaces for cultural life, water management,

Toward a Public Server Farm. Collage by Niklas Maak and Stefan Sauter.

environmental pollution, support for small shops and workshops, and local production." Citizens could share personal data with their neighbors without government oversight, marking a stark contrast to both China's model, where data deprivatization primarily serves state surveillance, and the US approach, where large tech corporations harvest behavioral data to create highly profitable products.

Could a third path—one that preserves technological sovereignty for citizens and allows them to retain the value of their own data locally—be reflected in a new architectural vision for data centers?

At the same time, one could imagine a Centre Pompidou of the digital age—a public server farm where the collective treasure of data is celebrated and displayed, much like the glass vitrines that modern architecture uses to showcase the art and consumer goods of its time. This New National Gallery for Servers would serve as a space for public education, revealing who controls artificial intelligence, platforms, the cloud, and algorithms; how we are analyzed, predicted, and monitored; what data sovereignty truly means; what happens when we lose control over our data; how we can reclaim it; and what new forms of participatory democracy might emerge. The avatars of individuals, housed within the data centers, could then become allies rather than mere digital reflections.

1. See "Modular Data Center Market Worth $79.49 billion by 2030 – Exclusive Report by MarketsandMarkets™," *PR Newswire*, November 5, 2024, https://www.prnewswire.com/news-releases/modular-data-center-market-worth-79-49-billion-by-2030--exclusive-report-by-marketsandmarkets-302296322.html.

2. See https://cloudscene.com/market/data-centers-in-united-states/all.

3. See "Großbrand in Datenzentrum sorgt für Ausfall von Websites," *Spiegel*, March 10, 2021, https://www.spiegel.de/netzwelt/web/ovh-grossbrand-in-datenzentrum-in-strassburg-sorgt-fuer-stoerungen-a-dff1fc32-8bd0-4305-a026-b6221e079455.

4. Rima Sabina Aouf, "AI's 'Eye-watering' Use of Resources Could Be a Hurdle to Achieving Climate Goals, Argue Experts," *Dezeen*, August 9, 2023, https://www.dezeen.com/2023/08/09/ai-resources-climate-environment-energy-aitopia/; and see Rima Sabina Aouf, "AI-Fuelled Data Centre Demand Will Set Back Energy Transition," *Dezeen*, April 25, 2025, https://www.dezeen.com/2025/04/25/ai-data-centre-demand-energy-transition/.

5. Paula Breukel, "Elon Musk: 80 Millionen Dollar für den Supercomputer Colossus," *Datacenter Insider*, April 1, 2025, https://www.datacenter-insider.de/elon-musk-80-millionen-dollar-fuer-den-supercomputer-colossus-a-72d264a169cb08aeb91875aa501bc389/.

6. Elon Musk held an official role in the Trump administration during 2025 as a Special Government Employee leading the Department of Government Efficiency (DOGE). This department was created by executive order at the start of Trump's second term, with the mission of reducing wasteful government spending and streamlining federal operations. Musk's tenure began early in 2025 and concluded around the end of May, as his 130-day term expired.

7. Ashley Belanger, "Thermal Imaging Shows xAI Lied about Supercomputer Pollution," *Ars Technica*, April 25, 2025, https://arstechnica.com/tech-policy/2025/04/elon-musks-xai-accused-of-lying-to-black-communities-about-harmful-pollution/.

8. Laura Larcan, "Elon Musk al Colosseo, tour speciale del monumento: 'Un grande appassionato di storia,'" *Il Messaggero*, June 16, 2023, https://www.ilmessaggero.it/roma/centro_storico/elon_musk_colosseo_visita_roma-7465912.html.

9. Karen Krüger, "Musk und Zuckerberg: Prügelei im antiken Theater," *Frankfurter Allgemeine Zeitung*, August 14, 2023, https://www.faz.net/aktuell/feuilleton/medien-und-film/musk-und-zuckerberg-wollen-in-antiker-arena-kaempfen-19100240.html.

10. Inga Janović, "Rechenzentren als Heizung," *Frankfurter Allgemeine Zeitung*, April 26, 2021, https://www.faz.net/aktuell/rhein-main/energiepolitik-wie-rechenzentren-als-heizung-fungieren-soll-17313513.html.

11. See "Management von CO_2e-Emissionen in hybriden IT-Umgebungen," Bitkom, https://www.bitkom.org/sites/main/files/2023-11/091123-Bitkom-Leitfaden-Management-von-CO2e-Emissionen-in-hybriden-IT-Umgebungen.pdf.

12. Siegfried Gendries, "Rechenzentren und Wasser: Was Datennutzung mit Wasserkonflikten zu tun hat," *Lebensraum Wasser: Der Wasser-Blog*, April 24, 2023, https://www.lebensraumwasser.com/rechenzentren-und-wasser-was-datennutzung-mit-wasserkonflikten-zu-tun-hat/.

13. See Scott McCartney, *Eniac: The Triumphs and Tragedies of the World's First Computer* (London: Walker and Co., 1999).

14. Martin Schmitt, "Die Geschichte des Potsdamer Rechenzentrums: Sozialistische Computernutzung und die Digitalisierung in Ostdeutschland," Lernort-garnisonkirche.com, June 6, 2020, https://lernort-garnisonkirche.de/die-geschichte-des-potsdamer-rechenzentrums-als-ort-sozialistischer-wie-demokratischer-verwaltungsautomation/.

15. See Niklas Maak, *Servermanifest: Architektur der Aufklärung: Data Center als Politikmaschinen* (Berlin: Hatje Cantz, 2022).

16. Evgeny Morozov, "Santiago Boys," *Internazionale*, September 28, 2023, https://www.internazionale.it/magazine/evgeny-morozov/2023/09/28/santiago-boys.

17. See "Facebook: Datenmissbrauch um Cambridge Analytica," Mainzer-Medieninstitut, https://www.mainzer-medieninstitut.de/facebook-datenmissbrauch-um-cambridge-analytica/.

18. See "Amazon knackt erstmals 2-Billionen-Dollar-Marke," *Manager Magazin*, https://www.manager-magazin.de/unternehmen/tech/amazon-onlineriese-erstmals-mehr-als-2-billionen-us-dollar-wert-a-f683a10b-1cf4-4b79-a454-5d2c9f6f607c.

19. See "AWS on JWCC for U.S. Defense," https://aws.amazon.com/de/federal/defense/jwcc/.

20. See https://old.snohetta.com/projects/388-the-spark.

21. See the architectural firm's website, https://www.schneider-schumacher.com/projects/project-details/828-qianhai-telecommunication-center/.

22. Dan Swinhoe, "Switch Plans Second Expansion of Pyramid Data Center Campus in Grand Rapids, Michigan," *Datacenter Dynamics*, November 6, 2024, https://www.datacenterdynamics.com/en/news/switch-plans-second-expansion-of-pyramid-data-center-campus-in-grand-rapids-michigan.

23. Francesca Bria and Evgeny Morozov, "Daten für die solidarische Stadt: Was wir gegen die neoliberale Smart-City-Agenda tun sollten," *OXI: Wirtschaft anders denken*, December 4, 2017, https://oxiblog.de/unsere-staedte-was-gegen-die-neoliberale-smart-city-agenda-zu-tun-ist-solidarische-stadt-daten-morozow-bria/.

24. Francesca Bria and Niklas Maak, "Holt euch eure Daten zurück!," *Frankfurter Allgemeine Zeitung*, October 19, 2020, www.faz.net/aktuell/feuilleton/francesca-bria-im-interview-holt-euch-eure-daten-zurueck-17007960.html.

Giulia Bruno

Leibniz Supercomputing Centre
of the Bavarian Academy of Sciences and Humanities

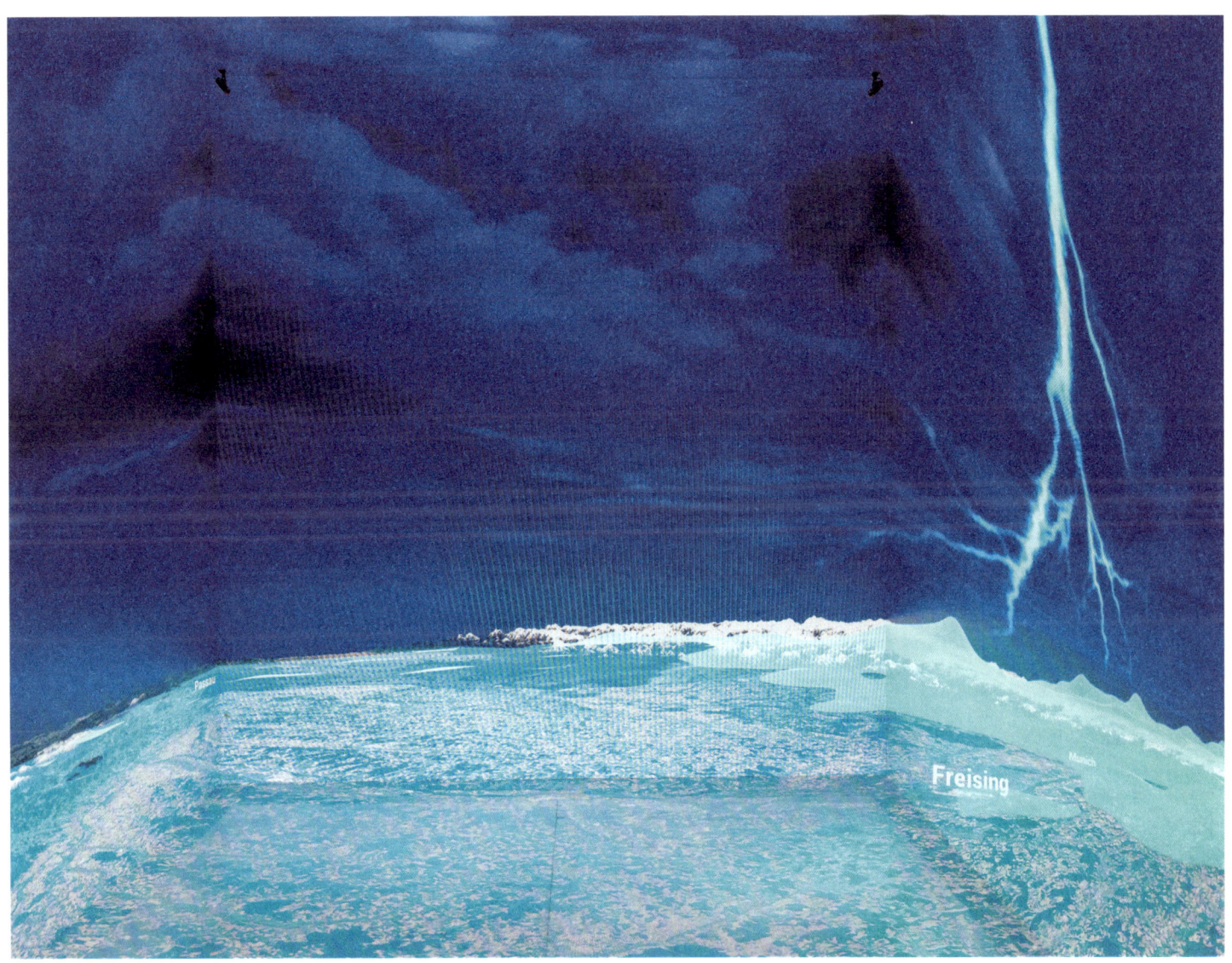

Extreme rainfall visualization from the ClimEx project (Prof. Dr. Ralf Ludwig, LMU Munich; ClimEx.org). 3D-immersive simulation based on the 1999 "Pfingsthochwasser" flood in Bavaria, used to communicate the impact of extreme weather events compared to traditional 2D mapping.

Spatial

Leibniz Supercomputing Centre of the Bavarian Academy of Sciences and Humanities

From the city center, after a short subway ride where the landscape rushes by first frantic and fast, then more expansive, you arrive on the outskirts of Munich, where the Leibniz Supercomputing Centre (LRZ) rises like a fortress of contemporary computation.

We enter a territory that occupies a material space, one that may seem abstract, but one that has its own shape and defined balance.

At times, it's easy to get lost imagining how it connects within an internal system, then interconnects on a global scale, and then becomes multiconnected once again in a service role—public and private. A territory with its own modules, micro and macro, which extend into others, traveling through kilometers of cables that mark paths and enable exchanges. New routes and constant returns. Cables that then traverse places, suburbs, cities, continents, and oceans. Infrastructures that, over time, become traces of economic, geopolitical, and cultural readings, and of power relations. The traded resource: data.

The supercomputer lives enclosed within an even vaster machine: a network of cables, connections, infrastructures, pipes, chips, colors, and an intermittence of light and dark, of cooling cells and control rooms that ensure its survival. The relationship between computational power and the infrastructure that enables it is overwhelming. It resembles a hard drive or a cube, a temple-like space. Like a hyper-technological organism, the center depends on a delicate balance: for every operation performed in milliseconds, a continuous supply of energy is required, then transformed, and eventually dissipated as heat. The cooling system moves labyrinthically, both hidden and revealed, like living tissue, and works tirelessly to maintain this balance.

The calculations made here represent an acceleration of time and a materialization of spaces once thought to be immaterial or intangible. In the silence, sound is reduced to a constant vibration, a soundscape made of space, fans, airflows, and electric impulses. The chips appear as precious structures, like gold ingots or portals to spatial worlds. It is a hard-to-define smell, bitter and sharp, reminiscent of heated plastic and freshly welded metal, artificial, a combination of warm air and a technical world where matter bends to computation.

The LRZ is not just a machine. It is also public architecture, an accessible, legible space where knowledge becomes service.

One wanders through corridors, floors, and spaces where diffused light gives way to darkness, and the only constant is a shrill background hum. The cables snake through every floor like roots, forming colored lines that to an untrained eye appear as an indecipherable mass. A fabric permeates both visible and invisible space, right up to the heart of the machines, which occupy space, create narrow corridors, and form parallel symmetrical lines.

Then, there is the empty space waiting for the future, to host new territories yet to come. The Faraday cage wraps the inner and outer wall to shield data and electromagnetic fields, but it also resembles a map whose grid marks lines and heights. It seems empty, and the organism seems to live a life of its own, like an aquarium that continues to communicate in code as you move through it.

Behind the presumed autonomy, human labor enables the functioning of one of the most powerful artificial brains in Europe.

Thus, the LRZ is both monument and machine, temple and office, technical and human space. It is a place where digital presence condenses and where the heat of artificial intelligence can only exist through the constant care of human labor.

Wafer Scale Engine 2 (WSE-2) from the Cerebras CS-2 system, the largest computer chip currently producible. Containing approximately 850,000 cores on a single silicon wafer, the WSE-2 is optimized for AI workloads, including large language models. At LRZ, it is used in research on applications such as automated hate speech detection.

Spatial

Legacy server room at LRZ with partially empty racks. Once populated with classic air-cooled server systems, the shelves now hold minimal hardware—mainly spare parts—reflecting the ongoing shift toward newer, high-density, and water-cooled computing infrastructures.

Legacy air-cooled server cabinets at LRZ, with partial occupancy by early-generation NVIDIA DGX systems. The DGX units, recognizable by their distinctive golden front inlets, were only available in air-cooled versions at the time. As a result, each cabinet holds just two to three systems due to thermal constraints. Most racks are now empty, reflecting LRZ's transition toward water-cooled, energy-efficient architectures.

Faraday cage structure within the outermost ring of the LRZ data center's "onion" architecture. This shielded corridor forms part of the infrastructure protecting sensitive computing systems—such as SuperMUC-NG—from electromagnetic interference, ensuring stable and secure operation of high-performance computing environments.

Backside of an active server rack at LRZ, showing dense network and power cabling. This tangle of data and power lines reflects the complexity of maintaining connectivity, performance, and redundancy in high-performance computing environments. Cable management is critical to airflow, serviceability, and system stability.

Spatial

Standard telephone installed within a network rack at LRZ, part of the center's operational infrastructure. While integrated among cables and digital hardware, these regular phones are present in every room of the computer center to ensure reliable communication, including direct calls to emergency services such as the fire brigade.

Argon-based fire suppression system within the infrastructure ring of the LRZ data center. Used primarily to protect backup and archival storage areas, the system ensures data integrity in case of fire without damaging sensitive electronic equipment.

View of SuperMUC-NG from the outermost ring of the LRZ data center's "onion" architecture. This perspective reveals the layout and infrastructure of one of Germany's most powerful supercomputers, supporting research in astrophysics, environmental sciences, and life sciences through large-scale simulations and high-performance computing.

SuperMUC-NG Phase 1 and 2, high-performance computing system at LRZ. Comprising over 364,000 cores and delivering more than 55 PFlop/s of combined computing power, SuperMUC-NG supports large-scale simulations across domain sciences including astrophysics, particle physics, computational fluid dynamics, and environmental and life sciences. Phase 1 includes 311,040 cores and direct hot-water cooling with waste heat recovery; Phase 2 adds 53,760 cores and was approved for full production deployment in June 2025.

"Warm aisle" within the SuperMUC-NG installation at LRZ. While most of the system is water-cooled, some components—such as switches—still require air cooling. To manage this, server racks expelling warm air are arranged face-to-face beneath a glass ceiling that contains and channels the heat, preventing it from dissipating into the wider data hall.

SuperMUC-NG Phase 2 and AI-focused computing modules at LRZ. Part of the next-generation high-performance computing infrastructure, SuperMUC-NG Phase 2 supports scientific simulation at scale, while adjacent AI systems accelerate machine learning workloads. This convergence of HPC and AI reflects LRZ's expanding role in enabling interdisciplinary research.

Spatial

Vacated section of the LRZ data center formerly occupied by the SuperMUC (Phase 1) system. The area is now being prepared to host Blue Lion, the next-generation successor to SuperMUC-NG. The upcoming system will require infrastructure upgrades to meet increased demands for electrical power and waste heat dissipation.

Photograph of the Cray Y-MP supercomputer, the first supercomputer installed at the LRZ, 1989–90. Initially delivered as a Cray-1 system, it was successively upgraded between 1991 and 1993 to 8 processors, 128 MB of main memory, and 80 GB of disk storage, making it the most powerful vector computer in the German academic sector at the time.

Q-Exa quantum computer installed in a production environment at LRZ. Developed by German-Finnish company IQM Quantum Computers, the system operates with a twenty-qubit superconducting processor cooled near absolute zero. In 2024, Q-Exa became the first quantum computer to be integrated with LRZ's flagship supercomputer, SuperMUC-NG, enabling research into hybrid quantum–classical computing architectures.

Linux Cluster poster at the LRZ, showing earlier Tier-2 high-performance computing infrastructure. Initially used by the ClimEx project for climate data simulations before transitioning to the Tier-0 SuperMUC-NG system, the Linux Cluster remains accessible to researchers from Bavarian universities.

SuperMUC-NG Phase 2, currently in acceptance testing at LRZ. This extension of the LRZ flagship supercomputer consists of an Intel Lenovo NeXtScale ThinkSystem cluster with 53,760 cores, approximately 28 PFlop/s computing power, 245 TB RAM, and 70 PB storage. The system was approved for full production deployment in June 2025.

Overhead InfiniBand network cabling of the SuperMUC-NG supercomputer at LRZ. Due to the water-cooling infrastructure supplying the system from below, the high-speed InfiniBand interconnect—responsible for linking all individual compute nodes—was routed from above, making it an architectural particularity within the data center.

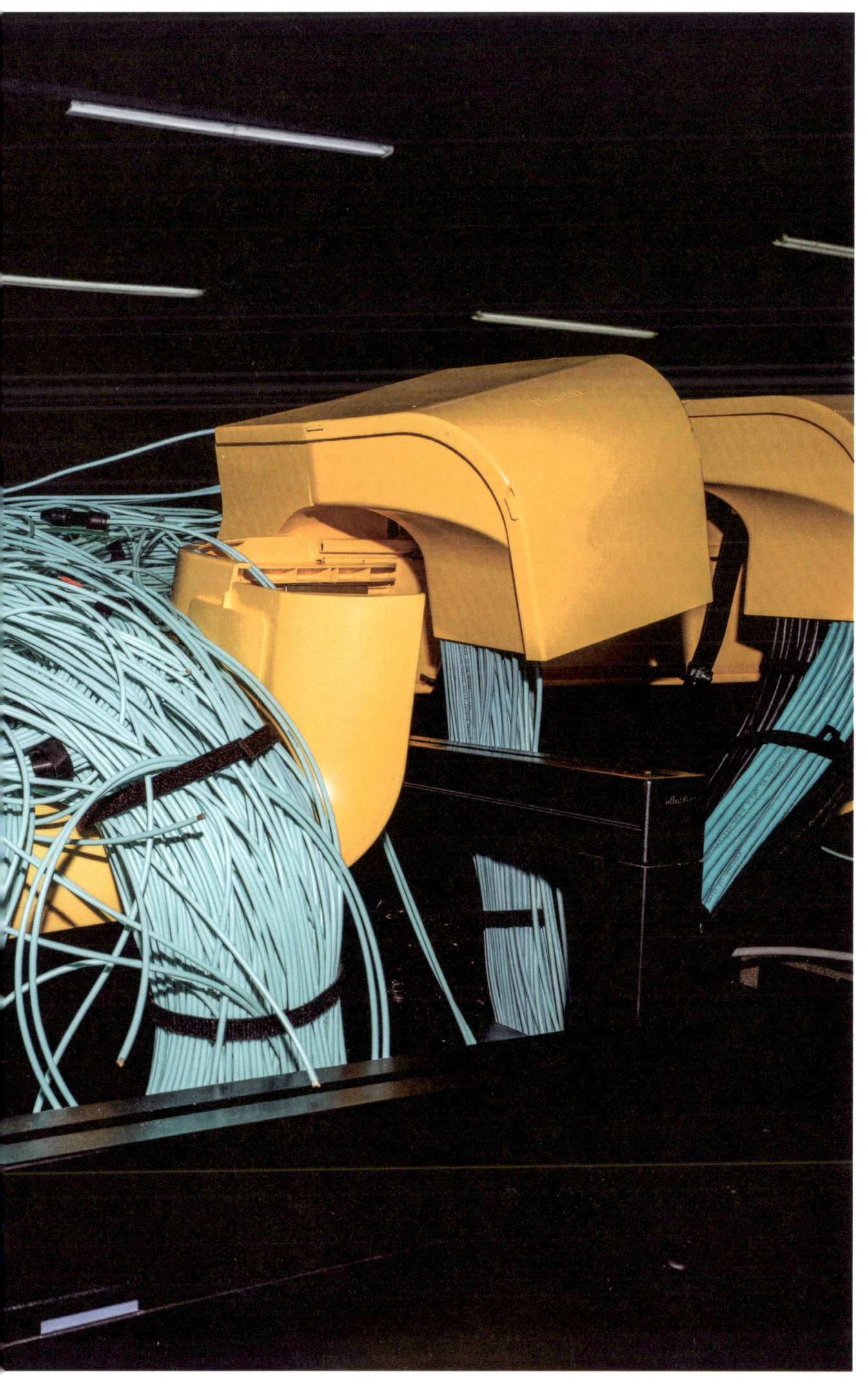

Undoing Optimization: Quiet Rewilding

The Future City

In 2006, I was interviewed by filmmakers about my visions for future cities. I had been attending the very nerdy Wizards of OS (Operating Systems) conference in Berlin.[1] These conferences, attended by thousands of programmers, artists, scientists, and advocates, investigated the role of technologies like free software in transformations of culture and politics toward versions that were more free, open, and democratic.

I stood in the sun and told the filmmakers about my vision of a home, a neighborhood where low-key infrastructure like wind and solar energy powered small, connected information boxes. I talked about how these could be homemade and maintained. I digressed into descriptions of craft, of neighborliness. I confess I imagined this as a quiet place, and technology as muted, intimate, not central.

The filmmakers were surprised at my answer to their question, and I can recall watching their faces shading into confusion. What great innovations would we have, they asked? How would the exciting new technologies that we were discussing as bringing democracy, transcending the conventions of law, and increasing the speed of innovation—how would these transform my future city, make it greater, faster, and smarter? Even for this progressive conference, the vision of technology was one of epic transformation.

What Is Technology For?

I had no answer for them, because (although I did not know it then) I was answering a different question. I had been wondering how people used technology projects to imagine their relationships—how the idea of open-source software production, featuring knowledge-sharing among an elite, shapes ideas about community or collaboration, and how these spread out into other settings like the creation of urban spaces and communities. First, I saw a vision of technology as democratic, but this shifted quickly toward new forms of control. The liberating feeling of posting pictures online to speak truth to power transformed into the menacing banality of Ring doorbells recording passersby in anticipation of sending the information to law enforcement. Such "smart city" technology creates optimal systems that can easily record transgressive actions and create heavily surveilled and managed spaces, like the "ring of steel" combining surveillance technology and hostile architecture in the City of London.

Across these supposed smart cities, computerized systems optimize things like spending on waste management, efficient policing, revenue collection through automated vehicle fines, and other data-driven aims that often undermine the experiences of residents, especially anyone marginalized by, for example, not having a place to live.

Aiming toward these aspects of optimization produces a fantasy of control, clarity, and efficiency within the bounds of the city. As my colleague Myria Georgiou points out in her book *Being Human in Digital Cities,* the conception of most smart cities has displaced the experience of many people in society—unless they are in positions of authority. This displacement suggests that optimized cities can transcend space and time and directs attention away from how surveillance and exclusion are not only associated with digital technologies—and how equipping cities with more of them has exacerbated existing inequalities and meant that more people experience pervasive surveillance,

Surveillance technology in the City of London.

which has historically been part of the experience of incarcerated people, migrants, and people in poverty. As visions of digitization and smartness evolve across global contexts, a technology-centric vision of optimized urban systems and flows can present the potential to replace, displace, or disregard existing modes of urban life. The popular narrative of smart cities (what Georgiou calls "popular humanism") here assumes that the introduction of digital technologies transforms cities into new entities—smart cities—made similar or comparable through their technological equipment. This ignores the other kinds of critical humanism that Georgiou identifies as part of urban life, including the many kinds of urban knowledge that don't exist in relation to "smart" technology as it is currently marketed.[2] This technology-centric urban vision draws on universal rather than pluriversal knowledge.[3] It also assumes managerial rather than communal values, creating not only a surveilled city but a unitary, rule-based, hierarchical one—a machine within the city walls.

Rewilding

Meanwhile, nonurban spaces are encouraged to "rewild." Experiments like the reintroduction of native species into landscapes such as the Knepp Estate in the UK, documented by Isabella Tree in her book *Wilding,* have transformed geographies and relationships. The return of species such as birds and butterflies to Knepp also inspired human visitors to spend time in the landscape and to rewild themselves. Tree writes, "Modern life, loaded with stimuli, multiple forms of communication and information requiring rapid processing and selection, demands … 'directed attention' … the natural

Spatial

environment, on the other hand, holds our attention indirectly, providing … 'soft fascination.'"[4] The book provides examples of visitors experiencing a sense of security when wandering the open wood pasture with its standing trees and open vistas.

Similarly, nature advocate Gina Maffey describes her own "rewilding" through a year spent living outdoors, creating different relationships with weather, seasons, and animals and moving from experiencing nature as "out there" and toward an experience of being part of nature. Her education and advocacy focus on connecting storytelling and lived experience of nature and the outdoors. Rewilding becomes a process of reintegration between people, their stories, and landscape that contrasts with the popular humanism of the smart city.

But something jars me. Rewilding often seems to resist the urban even though most humans live in cities, along with other creatures. This suggests a continued effort to maintain the distinction between nature and culture that's often characterized as part of the Enlightenment project. Optimizing within and rewilding without maintains both the separation between nature and culture and a cybernetic paradigm of continuous improvement that justifies intensified datafication and surveillance. Yet, this separation is untenable: life strives and even proliferates even under extreme situations, and the sprawling (and green) metropolis of London is declared more biodiverse than the surrounding countryside.

The concern then is not only the protection of the creatures who live among the humans of the city, nor even the encouragement of the kinds of physical environments that might allow all of these to physically thrive: more trees or bogs, clean waterways, wildflowers. It's also the invitation to connect "soft fascination" with focused attention—to see the connections between humans and their environment not as if the city were separate from nature, but as if it were, as it is, a part of nature.

Developing Quiet Rewilding within Cities

This quiet rewilding has three core components: observing things otherwise, making space for others, and sitting with difference. Each of these components connects with principles of Georgiou's "critical humanism," such as the importance of creating pluriversal experiences where many worlds (of knowledge, experience, or subjectivity) can fit, and of considering places on their own terms, not in comparison with others nor as representative of some singular "global" quality.

Importantly for me, this approach to quietly rewilding within cities doesn't reject technology outright (although it's critical of modernity). Nor does it overturn human "rights to the city" through an appeal to multispecies sensitivity. Instead, like my poorly understood vision of the urban future, this set of orientations turns toward the deeper questions about connection, attention, and flourishing within, between, and beyond differences. A community garden, for example, can help to grow food and habitats, as well as trust between the people who tend it.

Observing Things Otherwise

When I wrote the book *Undoing Optimization*, I was trying to explore what potential remained within the technology-driven "smart city" for perceiving urban relations otherwise. I thought about the way that digitally mapping living things like trees might provide humans a means to imagine how other species perceive the city—perhaps as a patchwork of snack shelters (trees) or a thread of life in a river? Or the ways that infrared burglar sensors triggered by the movement of foxes could be interpreted not as a malfunctioning human sensor but as accidental fox sensors, allowing a view of how a nocturnal creature might experience the city.[5]

With time I realized that these somewhat clumsy exercises rested mainly on my combining the "seeing" through data and sensor representations with the soft fascination or speculation that I usually bring to birdwatching or daydreaming along the river, something I've done my whole life, inside and far away from cities. As I learned more, I realized that the trick I was trying to pull was to enact a form of "Two-Eyed Seeing," combining a rationalist view of the world with a more embodied one.

Two-Eyed Seeing (*Etuaptmumk* in Mi'kmaw) is described by Elder Dr. Albert Marshall as one strategy to contest the erasure of traditional (sometimes called "non-Western") modes of knowing. As part of extractive and dominating dynamics, the collected and experiential knowledge of people about and within a place is often intentionally destroyed, erased, or dismissed. If this knowledge survives, it is often dismissed—it might be considered "primitive" or "traditional" in contrast to "up-to-date" or "modern." This erasure of knowledge and experience is described as epistemic violence, as a harm to ways of knowing.

Advocating for Two-Eyed Seeing in the support of fisheries, Elder Marshall and collaborators outline how some ways of seeing rivers (as living creatures, as relations) can be placed not as subservient to ways of seeing rivers as generators of resources (fish, water), but as equal to these.[6] The conflict between the two realities does not disappear in this practice. The fundamental incompatibility must be experienced, not only acknowledged. Declaring the existence of competing knowledges or realities (for example through land acknowledgment declarations) does not in itself address epistemic harm or violence. Agreements or actions based on Two-Eyed Seeing must contend with the reality that the mode of seeing that privileges rivers as resources has often underpinned violence wreaked upon people who see rivers as relations. These realities, once observed and experienced, can shape decisions about land and people in multiple directions, and not only strategies such as giving mountains or rivers "legal rights" but also including the reintroduction of responsibility for landscapes to traditional keepers, the renunciation of land ownership, or new principles of common access and traditional land practice. All of these are unfolding in places around the world, though not often in cities.

The exercise of seeing with two eyes, of attempting to imagine the experience of others, sits with difficulty alongside the experience of addressing how claims of rationality might also involve actions of dispossession.

Combining the experiences and perspectives of different denizens of the city opens up the kinds of potential and difficulty that Elder Marshall's Two-Eyed Seeing provokes. By looking for (or seeking to discover) the other experiences of the urban, it is possible to begin to see how cities create "others," and what is at stake when automation systems reinforce already dominant ways of seeing and knowing.

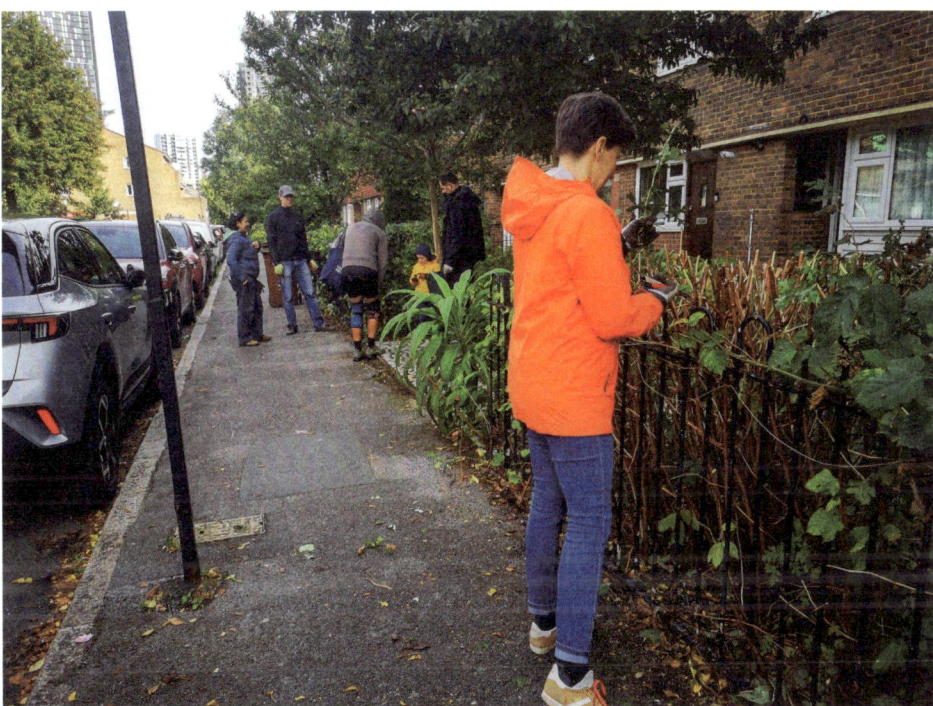

The author tends some blackberry vines with neighbors.

Making Space for Others

Years after her year of living outdoors, Gina Maffey and I joined a "motley crew" of artists, designers, philosophers, and scientists in a project called Rewilding the Night, led by philosopher of darkness Taylor Stone. Taylor, along with artist Rupert Griffiths, had been exploring ways to represent the level of light in cities at night—opening up conversations about the impact of artificial lighting on animals and plants. Rupert's "LightClocks" illustrate cycles of light and darkness in different places, from cities including Beijing and London to a nature reserve in Cumbria, North West England. The clocks help to visualize the cycles of light and dark across days, seasons, and years, spiraling from larger loops into smaller ones as time passes. Through the patterns of the light clock, we can see how bright days are in summer, and how dim, though never dark, London nights are, even in winter and without the moon. What's the significance of those dim, never-dark nights? Again, it's tempting to consider this in terms of contrast or opposition. In the dark sky reserve, Rupert might have stayed overnight among wildlife like hedgehogs, bats, badgers, or foxes, and might have experienced the heightened hearing sometimes associated with spending time in a dark place. In London, conversely, we might have expected brightly lit, "safe" streets—and little in the way of animals apart from domestic pets. In our research, though, we explored the complexity of the addition of artificial light to the city at night. Over dinner, the astronomer Hannah Dalgleish's voice wavered talking about the cultural fracture created by the fact that most humans alive today can't see more than a few stars in the sky at night. Iris Dijkstra, the lighting designer, described how bats perceive light, and how she had created a lighting scheme on a bridge that could change color temperature during the time that bats were flying through. We learned that hedgehogs can't navigate as well in twilight as they can in darkness, and why as a consequence they are disappearing from urban areas. Designer Nick Dunn told the history of darkness in cities as one of subversion, transgression, and danger, while also revealing that in past times, people carried their own lights with them while walking through dark streets, and that children were asked to run errands at night to accustom themselves to the dark. We all considered how expectations of brightness, visibility, and safety have driven the introduction of LED streetlights that operate at a higher frequency than the previous lighting, disrupting the foraging of hedgehogs and the sleep of humans.

Sitting with Difference

Darkness is not experienced in the same way for everyone. The bats who live somewhere near me in my inner London home awake with it, although they would likely prefer the nights to be darker, unlike the foxes whose ideal time is crepuscular twilight and dawn. My teenage daughter, raised in the city, might be worried about unexpected humans occupying dark corners, but not, perhaps, about foxes. The ability, capacity, and sensibility required to make space for these inhabitants, among their differences, is not easily acquired and takes effort to sustain. Priorities are set; resources are allocated. Claims are made about needs and rights. Bats, for example, are described by the UK minister for housing as preventing the building of houses for humans. In reality, it is relatively easy

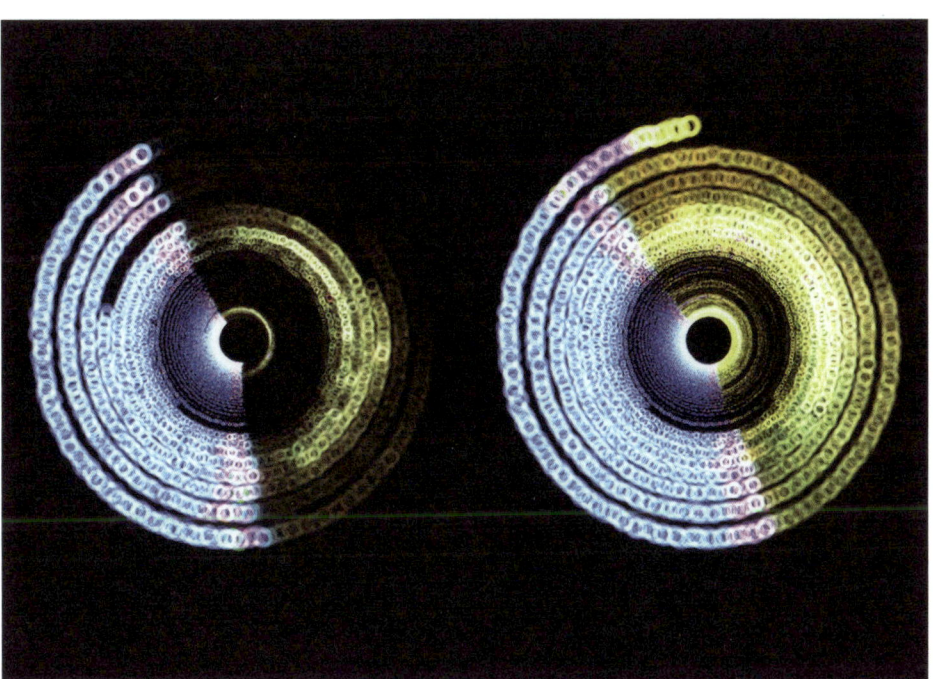

Time piece by Rupert Griffiths showing changing patterns of light and dark over a period of two months from December to January 2023/24 in Leighton Moss RSPB Nature Reserve (left) and Bonn Botanische Gärten (right).

to include bat habitat that doesn't impact humans in new housing design—less than 0.1 percent of building permissions in the UK are refused because of bat habitat infringement.

Among humans, who themselves have different experiences and needs, cities can continue to evolve as places to support and enable flourishing amid difference. It is Ramadan now, and I've hardly seen my next-door neighbors, as the season of fasting and reflection can be a quiet period. Soon though it will be Eid, and everyone in the family will be spilling in and out of their apartment, across the front path and into the backyard, where smoke will billow from the grill while feasts are prepared. Eid is not my festival. I listen to the singing and take in the joy filtered through the wall, making sure to close the upstairs window against the smoke. Once we've all gone to sleep, I'm sure the foxes will feast too.

A group of people dig an urban garden bed under the bare trees of late winter.

The pluriversal city is a reality, from the perspective of the many different humans striving for dignified lives to the bats, birds, foxes, cats, and bumblebees who also share and create a viable living environment. Because we are all in this together, sitting together is not only a matter of passive tolerance of difference but a striving toward facilitating the capacity for joy from our neighbors, the spaciousness of imagination that acknowledges that often another's flourishing contributes to ours too.

My quiet imagining of a technological future without obvious technology has broadened into a vision of a city that resists not the technology but the ideology of a controlling, dominating "smartness." No single transformation can be relied on to stabilize the future into being known or being predictable, not without losing so much of what it means to be alive, together, here and now.

Growing the Pluriverse

Many forests, biologists now understand, grow around complex networks of communication based on mycorrhizal growth of different species of fungus. These networks distribute energy between established and young trees of the same species, and between species. They send warnings of pathogens and electrical signals in times of drought. The fungi benefit too. Yet, until very recently, the presence of this network, never mind its complexity, was still in doubt. Understanding what it does helps to explain why many "isolated" trees across Europe are in poor health, and why tree planting in plantations does not bring as many ecological benefits as retaining ancient forests. The reductive tendency that sees forests as collections of trees and shrubs to be managed is mirrored by the tendency that sees cities as collections of functions to be automated. Against this tendency it might be useful to place the practices of Two-Eyed Seeing and the consideration of how the many kinds of difference such a complex organism or organization as a city can contain. This reiterates a pluriversal orientation toward the humanity—or the ecology—of the city. Humans have made cities, and because of this they are made for the benefit of humans. Like forests, though, which humans have also been part of, and shaped, cities are wild, and their diversity and complexity are part of what sustains and contributes to flourishing life on this planet.

1. The name is a play on words referring to the *Wizard of Oz*. For a short history of these conferences, see "Wizard of OS," Wikimedia Foundation, last updated July 17, 2021, https://en.wikipedia.org/wiki/Wizards_of_OS.

2. Critical humanism, for Georgiou, mobilizes power to enact more democratic thought and praxis within digital cities, pushing back against conceptions of technology that create hierarchies between humans. See Myria Georgiou, *Being Human in Digital Cities* (London: Polity, 2024).

3. A pluriverse is a "world where many worlds fit" according to anthropologist Arturo Escobar. See Escobar, *Designs for the Pluriverse: Radical Interdependence, Autonomy, and the Making of Worlds* (Durham, NC: Duke University Press, 2018).

4. Isabella Tree, *Wilding: The Return of Nature to a British Farm* (London: Picador, 2018), p. 218.

5. Alison B. Powell, *Undoing Optimization: Civic Action in Smart Cities* (New Haven: Yale University Press, 2021), Chapter 5.

6. Andrea Reid, L. E. Eckert, and J.-F. Lane, "'Two-Eyed Seeing': An Indigenous Framework to Transform Fisheries Research and Management," *Fish Fisheries* 22 (2021): 243–61, here 244, https://doi.org/10.1111/faf.12516.

7. See Rupert Griffiths, Taylor Stone, Alison Powell, Nick Dunn, Iris Dijkstra, Hannah Dalgleish, Luca Hector, and Andreas Müller, "Re-wilding the Night: Understanding How Darkness Is Valued through the Nighttime Light Ecology of Bonn Botanical Gardens," *Philosophy of the City Journal* 2 (2024): 12–29, https:/doi.org/10.21827/potcj.2.2.

Temporal

Marina Otero Verzier in Conversation with Cara Hähl-Pfeifer, Damjan Kokalevski, Andres Lepik, and Māra Starka

Temporal

The third part of the conversation addresses the responsibility to acknowledge the physical and environmental impact of rising digital footprints. Marina Otero Verzier outlines the ideas of "data loss" and "data mourning" to question whether an increase in information necessarily leads to more knowledge. She introduces data curation and ethical practices in digital archiving, along with the psychological implications of "data mourning." The conversation concludes with a proposal to integrate data curation into architecture education to bring awareness and critically address the relationship between technology and processes of knowledge creation.

Computational Compost prototype by Marina Otero Verzier and Donostia International Physics Center and film by Otero Verzier and Locument exhibited at the Tabakalera International Center for Contemporary Culture, San Sebastián.

Cara Hähl-Pfeifer, Damjan Kokalevski, Andres Lepik, and Māra Starka: The increase in data production leads to the exponential growth of our digital footprint. We need to take responsibility in recognizing its physical impact. Can we discuss practices of "data loss" as a critical consideration of this thesis?

Marina Otero Verzier: It's not necessarily the case that more information leads to more knowledge. I'll give you an example that comes from the realm of censorship and how intelligence agencies manipulate or obscure information. I had a conversation with a historian from the National Geospatial-Intelligence Agency in the US, and he told me that, at one point in history, certain knowledge was kept secret by preventing it from circulating by physically storing it in a secure location. Today, in contrast, we live in a moment of information overflow. There's so much circulating that it becomes difficult to tell what's true and what's false. In that context, certain narratives (even secret ones) are allowed to circulate freely, because in the middle of all that noise, it's almost impossible to discern which narrative actually holds.

So, we need to develop other ways of producing knowledge not based on the sheer quantity of information, but on our capacity to evaluate, analyze, and think critically. That's essential to our relationship with data proliferation and with AI. Especially now that we rely on infrastructures like artificial intelligence to provide us with information, the question becomes: How do we assess the value and accuracy of that information? We already know that an overflow of information can be overwhelming, even leading to a form of illiteracy, where we struggle to navigate or make sense of it all. And that's a paradigm we're still trying to navigate. It applies to many different areas, and it reinforces how vital critical analysis is if we want to make sense of this complexity.

Then there's the question: Why do we accumulate so much data, and who benefits from it? The obvious answer is companies like Google, Meta, or Amazon—they profit from that accumulation. But if we think beyond the corporate angle and instead consider the relationship between human knowledge and the amount of data we generate, I'd be curious to see how that curve actually correlates. Does it rise together with misinformation, disinformation, conflict, and violence?

We see similar patterns in scientific research. I've visited labs where enormous volumes of data are generated every day: astronomical observatories, particle physics labs like CERN, or facilities collecting information about neutrinos. In some cases, like certain telescopes, they can't even store all the data they capture in a single night. It's just too vast. AI is used to filter through it, flagging any variations from the night before. The hope is that, within that accumulation, something will trigger a breakthrough—a new equation, a paradigm shift, or a scientific law that reframes how we understand the world. Then, perhaps, we'll know what to look for, and how to gather it more selectively.

At the same time, this accumulation is happening in parallel with the cancellation of many possible futures due to climate catastrophe. We're generating vast amounts of data for a future that might never arrive, or that comes at the cost of other vital futurities.

> In the exhibition context, we often talk about data curation. Additionally, we are currently establishing a born-digital archive at the museum. So, it's a practice that will become increasingly more important for historians, in education, as well as architecture and design practitioners.

The question of data curation is very important, especially how we find information, how we assess it, and how we stay critical of the information we receive. It also relates to questions of selection and preservation: what we keep and what we let go. We're used to making those kinds of decisions in many other areas: through archival policies that determine what enters a collection and what doesn't, or through architectural, cultural, and legal frameworks that decide which buildings are preserved and which are not. But with data, we haven't yet developed a clear framework or practice. That's partly because we've been living under the illusion of unlimited digital space, this idea that we don't have to let anything go.

And that's a problem. First of all, it's not true that digital space is unlimited. Data decays—it's actually less stable and more fragile than paper. Secondly, digital infrastructures are entirely material, and loss is already embedded in the process of storing information. Files are constantly being updated, migrated, mirrored, reformatted, reshaped through cycles of technological obsolescence and regeneration. So, the real question is whether this massive accumulation and flow of data, which depends on algorithms to access and analyze information and often ends up feeding back into itself, is affecting our capacity to think critically and generate new knowledge. That's why I believe we need to retain more agency in the process through a renewed focus on data curation. It compels us to confront the question of limits, even within the supposedly immaterial realm of the digital.

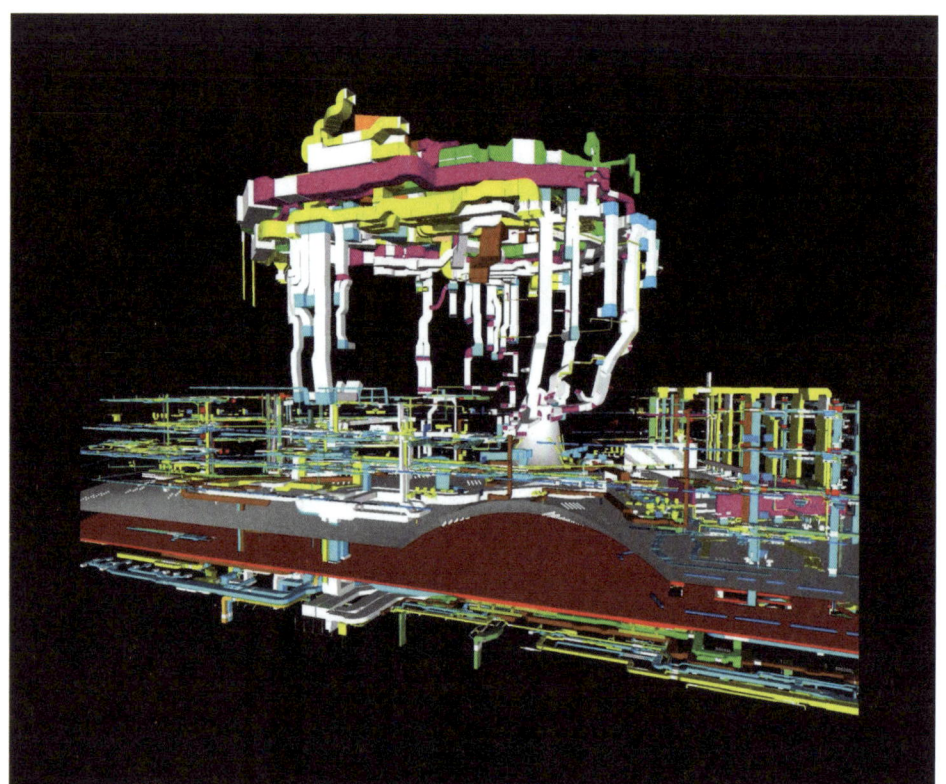

Technical building services digital model of the Elbphilharmonie Hamburg, designed by Herzog & de Meuron, 2017.

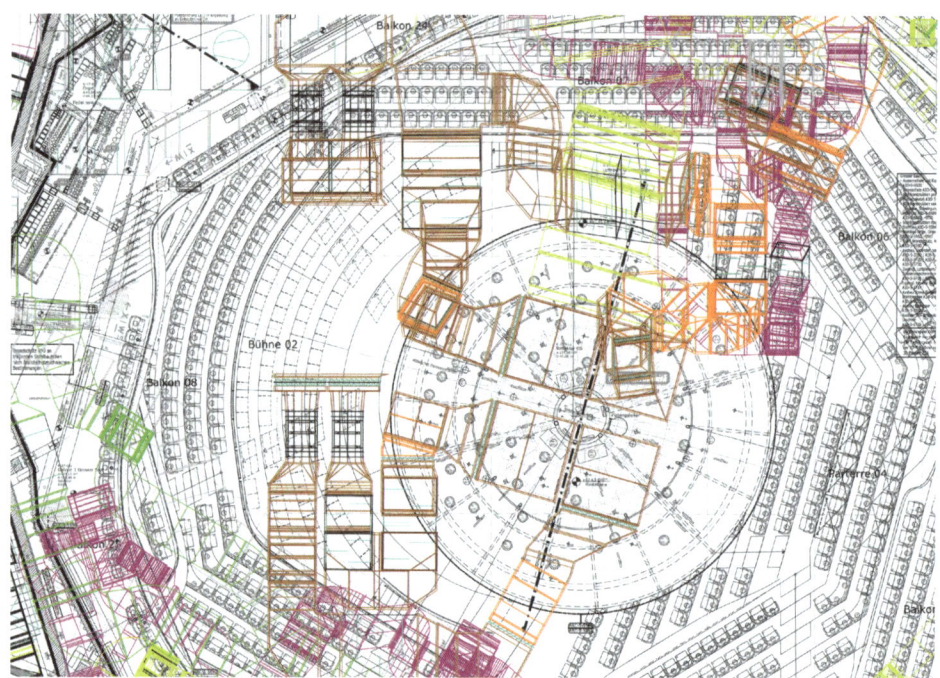

A zoom-in of the Level 17 floor plan showing the mechanical ductwork of the main concert hall at the Elbphilharmonie Hamburg, designed by Herzog & de Meuron, 2017.

In my work, I often speak about data mourning. Any form of memory-making is always a balance between remembering and forgetting. That's inevitable. I use the term "mourning" because it brings in the notion of consciously letting go. That's not the same as forgetting; it's more like remembering differently, through a purposeful act. I also use it because I think there's something psychological at play here, a kind of compulsion to accumulate data.

Mourning, in this sense, recognizes the emotional and psychological dimensions of our relationship to information and the quest for more data, which is not driven only by function or profit. It's also an obsession.

Data curation can be a risky endeavor in light of the current US government pulling and deleting valuable historical data from governmental websites offline, thereby censoring access to what should be public information. It's highly troubling seeing all of that spinning out of control. Funnily enough, so-called data rescuers are now attempting to preserve and back up this endangered data.

That's a great point, because while there's so much talk about data accumulation, how our data is stored—and by whom—also carries the risk of it being misused or deleted. I'm always cautious when I advocate for data mourning, because the loss of certain data holds strong political implications, especially in countries that have endured, for instance, a dictatorship—where there has been a systematic erasure of certain parts of the population and their histories for decades. And that's exactly what we're seeing now. So much digital information is stored in data centers we don't have access to—managed by a handful of actors—who can, at any moment, delete entire datasets containing valuable information. The mission of companies like Google or Meta isn't to preserve information against erasure, but to monetize it. That's why initiatives like Data Refuge projects, shadow archives, and groups such as Indigenous Data Sovereignty are so important—they actively resist this dynamic and work to safeguard knowledge that might otherwise disappear or be misappropriated.

> There is an immense responsibility when data is curated and archived. We're shaping certain narratives of history for future generations. Exactly how, to this day, the ideal of white, male-dominated architecture of the twentieth century is perpetuated in teaching. Do you think data curation should be more integrated into education for future architects?

I think it's the only way forward now. Because, for instance, when you look for information on a topic today, how do you decide what's relevant when you're searching on Google? Already, the first results are AI-generated. Some are sponsored or promoted by

Oceanic Sovereignty: Beyond the Coordinates of Loss "Alternative Maps as Testimonies, not Territories." Student work by Viksha Nayak, Vishesh Sahni of the 2025 Data Mourning Clinic led by Marina Otero Verzier and Dan Miller at Columbia GSAPP.

the algorithm. So, within that logic, teaching people to understand the mechanisms behind the platform—how to read it, how to scan it, and how to curate what they find—that becomes fundamental.

> Not just that, during their studies, young architecture students are producing a lot of data. Is it necessary? What is relevant to save? When you move architecture offices afterward, how do they preserve data? How do they select what to save? There's no selection process. And so, all those terabytes of data get gifted to the collections and become a question for archiving. In a way, it's a new problem. This born-digital material requires different archiving procedures than the ones we are used to. So, how do we archive, how do we teach, and how do we research in the future?

It's complicated. I was recently at the Canadian Centre for Architecture in Montreal, trying to look into a project by Philippe Rahm. He had used a specific software to create an installation, and while the files are fully preserved in the archive, the institution also had to save the software itself and ensure it remains operational on the hardware available, so that researchers can still access it. But while consulting the records, I realized I didn't know how to use the software. I was given access to everything—the folder, the files, the software—but I couldn't figure out how to operate the programs, so I ended up unable to research the installation through the archive. So, does that mean that, in the future, people working in archives will need to know BIM software, Blender, or AutoCAD? That kind of technical knowledge might disappear in thirty years, and that poses a real problem.

Preserving and accessing digital information for research is not without difficulties. First, there's the challenge of selecting from terabytes of data that

Temporal

Data Forest storage overview, part of the project *Cultivating Feminism[s]: A Cyborg Future for the Data Archive* by Helena Francis, 2021. The project explores advances in biotechnology to store archival data in the DNA of living plants and was developed within the ADS8 studio Data Matter: Digital Networks, Data Centres, and Post-Human Institutions at the Royal College of Art, led by Ippolito Pestellini, Marina Otero Verzier, and Kamil Dalkir.

are often practically illegible. Then comes the task of maintaining the necessary software and hardware. And finally, being able to actually run the programs. Many people are now trying to print their materials or take screenshots, turning information into more "banal" formats and file extensions. So the solution isn't just curation, it's also resolution and facilitation of access. Otherwise, the impossibility of reading these files will render them illegible and turn them into digital debris.

> Since a big part of your work is teaching, have you developed a certain guideline for student-led projects concerning medium, technology, and perhaps also quantity?

When we traveled to Tuvalu for the Data Mourning Clinic at GSAPP, we developed a series of protocols for engaging with digital technologies in fieldwork. For example, what happens when twenty students arrive on an island and start photographing the same thing? It has both social and ecological impacts. One of the key questions we asked ourselves was whether we should use a drone. In the end, we decided against it because it felt intrusive to have a drone flying over the island, capturing footage without the consent of the people living there. We wanted to ensure we had someone's consent when taking photos or recording interviews, so we approached this with care, following the IRB guidelines for ethical research practices. We also discussed how many photos were appropriate to take, as we wanted to avoid contributing to data accumulation.

Then there's the question of infrastructure. Some students in the class are building a server powered by solar energy. They've become experts in feminist servers and permacomputing movements and are working on developing a system that everyone in the class can use. We're using low-resolution images as part of this, turning it into both a political and aesthetic practice. It's a playful process, but one that also highlights the potential of using technology differently in architecture. I hope some of the students will come to appreciate the beauty of a rendering that's not fully polished but has intention behind it. We need to reflect on the usage protocols of these digital tools as something we can appropriate and transform. Education should recognize our reliance on these technologies and help us critically engage with them.

> And do you then have a specific protocol of how you would go ahead and archive the projects?

We're working on it. One outcome of the studio and seminar may be the development of a decentralized archive and exploring its stewardship. We've already discussed eliminating duplicates and consolidating everything to avoid individual silos. We're inspired by examples like the *Low-Tech Magazine*, whose server only functions when there's sunlight. There's a direct relationship between accessibility and environmental conditions.

We also believe the archive doesn't need to be accessible all the time. The other day, we discussed why digital spaces should operate 24/7 when most physical ones do not. Maybe digital spaces should have opening hours that align with the local time cycles of the places where they're accessed. Why not? It's a beautiful idea. We want to align more with natural cycles and avoid treating the digital as an isolated abstraction.

> Today, we see some architectural practices going back to hand drawings and physical modeling, a practice that was lost in many schools at the beginning of the century. Are we restricting the media of our profession too much?

Yes, but there's a reason for that. Many of our documents serve as legal instruments. They act as interfaces between multiple parties, which is why they adhere to specific formats. I'm talking about technical drawings, for example. These documents have to be part of a larger conversation and share common codes with others who understand them, and that's hard to change.

This is part of the work of a professional with legal responsibilities. Those responsibilities are often embodied in documents, contracts, and technical drawings. But as the scope of our profession grows, so do the mediums we use. For example, videos have become an essential tool for architectural offices to communicate ideas, and AI is dramatically shifting how people approach design. There are many other ways to engage with media. At the Royal College of Art, for example, media studies are integrated into the architecture department. I think this is fantastic because it exposes students to a variety of mediums while also diving into their histories, implications, and potential uses. I've noticed that the students are much more precise about how and why they choose a medium, as well as what it means to use it in a specific context.

> It sounds like you are advocating for returning to analogue business, analogue methods of producing and archiving architecture, and implementing them on the digital.

As I mentioned, the key questions revolve around resolution, accessibility to digital spaces, and collective knowledge as a common good. This means not pushing devices to their limits, as doing so increases energy consumption and heat production. As architects, when we render, we often push systems to their maximum capacity. That's why computers break down—it's always the day before a final presentation when someone says, "My computer crashed." We're constantly stressing systems to the breaking point, and when we do that, we're also pushing ourselves and others to the limit.

Renderings need to be understood in terms of the human labor involved and their social and environmental impact. Many architectural offices set up rendering farms in different time zones, outsourcing labor and seeking the cheapest, most vulnerable workers. This has been explored by Liam Young in his project *Renderlands*, where he argues that what we think of as visions of the future, designed by Western architectural offices, are actually produced by outsourced render farm workers in the Global South.

We're exploring alternatives to these exploitative practices: low-speed approaches like underclocking, which are part of what the permacomputing community advocates. These ideas really interest me. Perhaps the future will be slower and more low-resolution than we imagined.

Andra Pop-Jurj

Constructing Cartographies:
Mapping Territories through Time and Technology

This visual essay examines the role of data and cartography in territorial disputes, tracing their evolution from colonial flag-planting rituals to contemporary digital mapping. It positions mapping as a visual form of knowledge production at the intersection of digital heritage, environmental preservation, and destruction. Through an Arctic lens, it explores the technological turn—from early polar expeditions to remotely sensed and digitally reconstructed landscapes—interrogating the shifting forms and material traces of environmental data. By deconstructing cartographic exercises conducted as part of the project Monsters and Ghosts of the Far North, this piece illustrates how waters and territories above 60°N have become sites of intense geopolitical and infrastructural intrigue, with incompatible and interlocking border claims rooted in colonial and cartographic history.

Arctic Materialities

The Arctic is at once an "ambiguous materiality, a global warming ground zero, an Indigenous homeland, a potential Indigenous state, a nascent global governance regime, a coveted wildlife park, an oil bonanza frontier, and a throwback to colonial land-grabbing all in one."[1]

The ambiguous materiality as proposed above refers to the multiple manifestations of the Circumpolar North. Expanding on this idea, this study examines the physical yet fluid politics of water as it shifts between solid and liquid states. It considers the Arctic[2] as a contested architectural and infrastructural space, shaped by the ever-changing traces of environmental degradation—freezing, thawing, and flowing.

Drawing on/extending this notion of ambiguous materiality, I suggest we understand governance itself as a shifting current, spilling into maritime realms as the ice recedes. Simultaneously, the Arctic offers a unique lens on anthropogenic climate change, where transformation—rather than linear decline—defines the region's shifting states. With degradation occurring nearly four times faster than elsewhere, change itself is instrumentalized to justify new resource-extraction schemes and industrial expansion.

Arctic ice manifests as permafrost, glaciers, ice sheets, and sea ice, each playing a crucial role in shaping our planet's climate, geography, shipping corridors, resource extraction, and geopolitical dynamics. As temperatures rise and Arctic Sea ice declines at a rate of approximately 13 percent per decade, glaciers are retreating, ice shelves are collapsing, and permafrost soils are thawing. Once a seasonal, cyclical process of melting and freezing, these phenomena have now diminished steadily and are on track to disappear entirely.

The notion of contemporary nature as a hostile witness radically reconfigures the perception of the Arctic not as an inert, frozen expanse, but as an environment once intimately navigable and legible to those attuned to its rhythms. Where ecological registration systems once guided the indigenous custodians of the land with precision, climatic aberrations now distort the contours of snow-laden terrains, rendering the Arctic increasingly unstable and unpredictable.[3] Matter is out of place, and the very landscapes that could once be trusted have become volatile.

Permafrost Metabolisms

Permafrost, defined as a thick subsurface layer of soil that remains frozen for at least two consecutive years, covers 22 percent of exposed terrains in the Northern Hemisphere.[4] This geological formation serves as a critical bridge between terrestrial and marine environments. It underlies vast stretches of Arctic tundra and extends beneath the seafloor, forming submarine permafrost that originated during the last Ice Age when sea levels were lower. This frozen ground acts as a structural and ecological link, influencing both land and sea systems. As permafrost thaws, it accelerates coastal erosion, reshaping shorelines and releasing stored carbon into the ocean and atmosphere.[5]

Permafrost cores have been drilled and sampled, unveiling ancient histories, yet much of their content remains a mystery. What lies buried deep within is not only a flash-frozen past but also a future. It is a silo of sorts, a granary of diamonds and ore, of fossils and gases, a bank of wonder and life. Stored in this vast basement of strange frozen treasures are millions of years' worth of species and eras of the tundra.[6] As this world melts, there is carbon and methane, mercury and smallpox, anthrax and Spanish flu, half sleeping in wait.[7] Thaw brings them one step closer to resurrection.

The accelerated transformations in the Arctic defy the notion of a frozen landmass, thus undermining the very principles of territorial integrity.[8] Its fluid materiality questions the possibility of a natural border and poses a need to imagine a more indeterminate idea of nature than the one constructed over the past centuries. The reality of a continually melting and moving Arctic challenges Western conceptions of territory and opens up the possibility of recasting and reclaiming uncertainty as a productive condition where such entanglements may be redefined.

Territories of Power

Mapping has long been a tool of territorialization, shaping how the Arctic has been understood, governed, and exploited. Early European expeditions, whether led by British, Danish, or Russian explorers, produced maps that not only documented but also actively constructed geopolitical imaginaries, often accompanied by colonial flag-planting rituals.[9] This process transformed the Arctic into a contested space continually redefined through cartographic techniques.

Beyond charting coastlines and asserting territorial claims, cartography has played a formative role in shaping the Arctic's urban and infrastructural development. While early maps served as instruments of colonial possession and geopolitical control, they also guided the planning of Arctic settlements, shaping infrastructure placement, resource distribution, and spatial hierarchies that embedded geopolitical agendas into the built environment. Today, Arctic data infrastructures—from satellite arrays and data centers to military bases and research stations—continue to mediate resource extraction, exploration, and surveillance, revealing the extent to which mapping is entangled not only with representation but with the production of territory itself.

Drifting ice stations, such as the Soviet-era North Pole series (1937–ongoing), represent an evolving approach to Arctic research, initially relying on natural ice floes and now transitioning to research vessels due to climate-induced ice loss. Sites such as Fletcher's Ice Island (T-3) served dual purposes of research and espionage during the Cold War while extending territorial logics of control far beyond the official confines and boundaries of the American state.[10] Even though the physical properties of an ice island challenged commonplace understandings of terrain, T-3 essentially functioned as an incubator where the research program became the technology of territory.[11] The twenty-five-year occupation of T-3 by the US ultimately demonstrates that along with infrastructural development, research into a place's physical geography is inherently territorializing.

These instruments—from scientific outposts to remote sensing technologies—not only document environmental processes, but also enact political influence, driving uneven development and urbanization across remote geographies. Marlene Laruelle's three-phase model—comprising the colonial, industrial, and globalization/circumpolar phase—provides a comprehensive framework for understanding this dynamic.[12] The colonial phase, spanning from the Middle

Ages in Russia and Scandinavia to the nineteenth century in North America, is reflected in early mapping practices[13] and territorial claims, as well as Danish efforts to formalize settlements and resource control through the establishment of trade, administrative, and missionary posts.[14] The industrial period, which unfolded primarily between 1928 and 1991, is embodied by Soviet Arctic urbanization during the mid-twentieth century, when rapid expansion of cities like Norilsk and Tiksi for resource extraction and militarization was underpinned by advanced mapping techniques and centralized planning.[15] Finally, the circumpolar era, driven by globalization, continues to reshape the region through evolving infrastructures and transnational data networks that complicate and reinforce overlapping territorial claims.

Cartographic Gaps and the Limitations of Territorial Representation

Despite advancements in Arctic mapping, particularly since the Cold War period, many conceptual and material/scientific gaps remain. The Mercator projection, still widely used in global cartography, distorts polar regions, compressing Arctic land masses while leaving central voids—areas that appear unmapped or empty despite their geopolitical significance. To address these distortions, the Universal Polar Stereographic (UPS)[16] was developed, providing a conformal projection for polar areas. However, while the UPS projection maintains angular relationships, it introduces increasing distortion as one moves away from the center, particularly beyond latitudes of 81°06'52" North or South. This paradox persists in contemporary digital mapping: while satellite imagery and GIS technologies offer unprecedented detail, they struggle to capture the fluidity of ice formations, seasonal shifts, and nuances of disappearing landforms, underscoring the ongoing challenges in accurately representing the dynamic Arctic environment.

During the Cold War, the Soviet Union employed advanced cartographic techniques, including sonar, seismic surveys, magnetometry, and gravimetry, to map the Arctic Ocean floor for both scientific and military objectives. These efforts, such as the mapping of the Lomonosov Ridge, have had a lasting impact, underpinning modern territorial claims in the region.[17] Building upon this legacy, contemporary cartographic research leverages satellite-derived data to analyze Arctic landscapes.

For instance, the ArcticDEM project utilizes high-resolution optical imagery from the Maxar satellite constellation to produce detailed digital elevation models (DEMs) of the Arctic, facilitating nuanced examinations of topographic changes over time. Additionally, the International Bathymetric Chart of the Arctic Ocean (IBCAO) provides comprehensive gridded bathymetric data, offering insights into the seafloor's depth and shape. These resources, among others, have become instrumental in modern Arctic research, enabling detailed analyses of both terrestrial and submarine features.

The Circumpolar Imaginary

The advent of satellite and GIS technologies has significantly enhanced our understanding and management of the environment, enabling a data-driven worldview. The proliferation of remote sensing instruments, coupled with increasing computational processing capacities, has fostered a circumpolar imaginary of the Arctic—one that meticulously tracks environmental shifts such as retreating ice sheets and glacial dynamics.

Advancements in computational power have transformed satellite remote sensing. Modern satellites now possess onboard high-performance computing capabilities, allowing for real-time data processing and analysis directly in orbit. This reduces the need to transmit vast amounts of raw data to ground stations, thereby decreasing latency and enhancing the efficiency of environmental monitoring.[18] Key to this advancement are platforms such as the Svalbard Satellite Station (SvalSat), NASA's Ice, Cloud, and Land Elevation Satellite-2 (ICESat-2), and the European Space Agency's (ESA) CryoSat-2 mission. Since the establishment of SvalSat in 1996 and more recently with the deployment of specialized instruments such as ICESat-2, satellite-based mapping has provided high-resolution, near-real-time data critical for conservation, preservation, and geopolitical strategy.

Established in 1997 near Longyearbyen at 78 degrees North, SvalSat's strategic location enables near-continuous communication with polar-orbiting satellites, facilitating frequent data downloads—up to fourteen times per day per satellite. The rapid relay of satellite data facilitated by SvalSat supports near real-time environmental monitoring critical for tracking sea ice changes, monitoring permafrost thaw and its impact on global climate feedback loops, observing atmospheric conditions, and supporting early warning systems for extreme weather events and environmental hazards.

SvalSat thus serves as a key geopolitical and scientific asset. Governed under the Svalbard Treaty, which grants equal access to signatory nations, the station supports multiple international space agencies and research institutions. This fosters a wider range of data collection and distribution, promoting both scientific collaboration and geopolitical engagement over Arctic environmental knowledge.

CryoSat-2 and ICESat-2 are complementary Earth observation satellites frequently used in tandem to study the cryosphere. CryoSat-2 (launched in 2010) uses synthetic aperture radar (SAR) interferometry to track changes in sea ice and ice sheets, while ICESat-2 (2018) employs photon-counting LiDAR to measure surface elevation and assess vegetation canopy height and carbon sequestration. Though not specifically designed for permafrost monitoring, their data—alongside advancements like Sentinel-1's SAR—have significantly improved the detection of ground deformation and ice thickness changes. However, despite these technological advancements, emissions from thawing permafrost are often underrepresented in global carbon budget models,[19] potentially leading to an underestimation of the emissions gap required to meet climate targets.

3D reconstruction of the Arctic Ocean floor based on DEM data.[20] The extent of higher latitudes pictured here corresponds to
The Arctic Council's Arctic Monitoring and Assessment Programme Working Group Boundary (AMAP), which incorporates elements
of the Arctic Circle, political boundaries, vegetation boundaries, permafrost limits, and major oceanographic features.[21]

Temporal

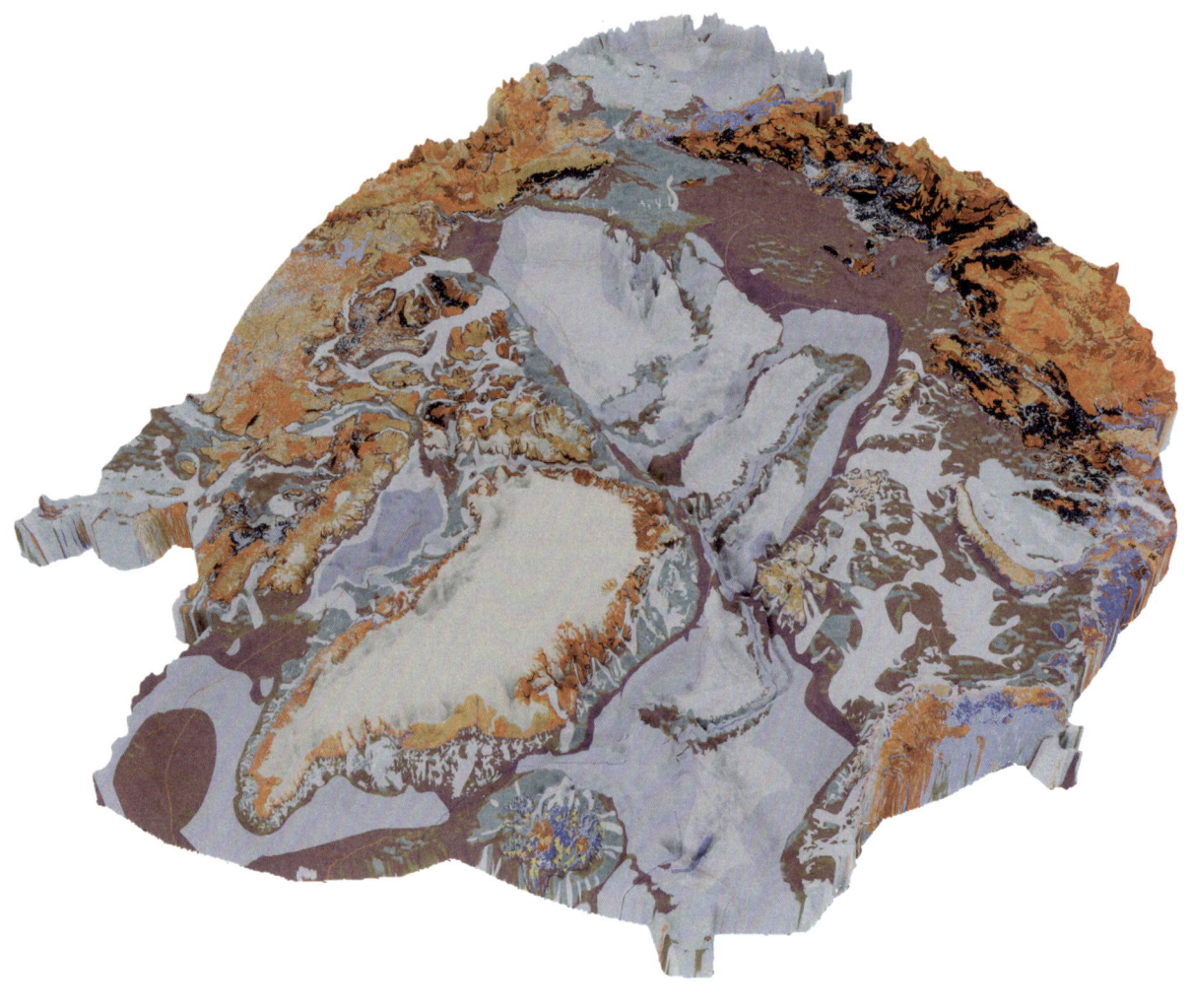

Permafrost landscapes constitute a spatial domain shaped by planetary exchanges of matter, offering a distinctive framework for interdisciplinary research into environmental degradation—ranging from microbial processes like methanogenesis to large-scale cryospheric landform dynamics. The map depicts permafrost soils, wetlands, and peatlands as prime habitats for methanogenic archaea. The exceptionally slow microbial metabolisms in permafrost soils offer unique transtemporal environmental insights while sustaining diverse methanogenic communities. With metabolic rates up to a thousand times slower than in temperate soils, permafrost preserves organic carbon over extended periods, providing a window into past ecological conditions.[22]

Vegetation Type
Source: Conservation of Arctic Flora and Fauna (CAFF)
Classification following Circumpolar Arctic Vegetation Map (CAVM) subzones
- Lagoon
- Wetland
- Glacier

Water Sources
Source: CAFF
Classification following CAVM subzones
- Lakes
- Rivers
- Reservoirs

Geomorphic Features Seafloor
Source: Geoscience Australia, GRID-Arendal, and Conservation International
- Ridges
- Glacial troughs
- Shelf valleys

Soil Carbon Database
Source: Bolin Centre for Climate Research; measured 10 cm under the ground
[in kg/m^2]
- 0–10
- 10–25
- 25–50
- 50–100

Carbon Collection Points
Source: Bolin Centre for Climate Research; from Northern Circumpolar Soil Carbon Database version 2 (NCSCDv2) pedons

Volcanoes
Source: United States Geological Survey (USGS) – arctic geology

Permafrost Extent
Source: Arctic Monitoring and Assessment Programme (AMAP)
- Continuous (90–100%)
- Discontinuous (50–90%)
- Sporadic (10–50%)
- Isolated (0–10%)

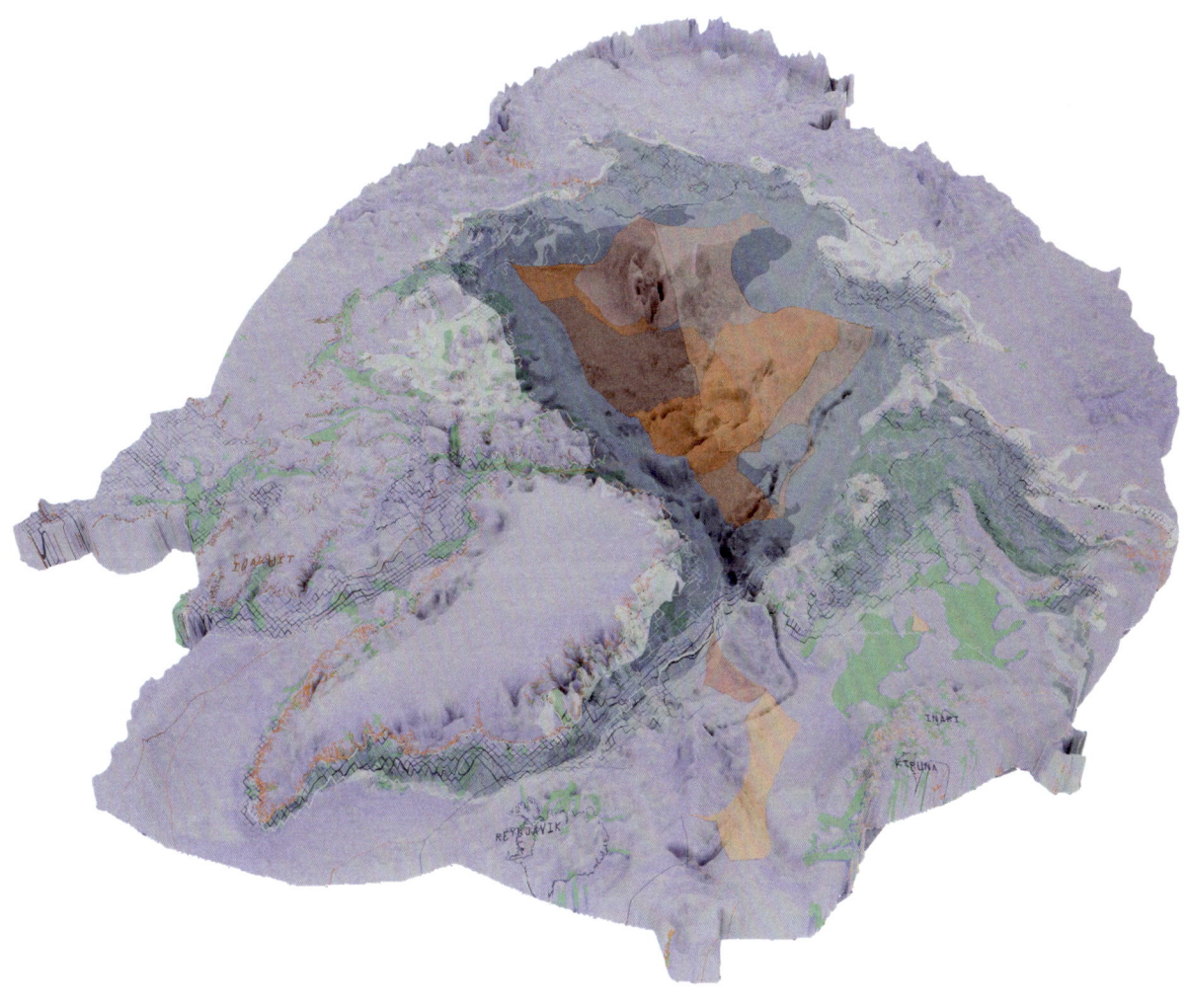

A key aspect of this process is the way in which environmental change has come to signify not only climate transformations but also shifts in geopolitical and economic dynamics. The retreat of sea ice, for instance, is directly linked to discussions of newly accessible shipping routes, resource exploitation, and territorial competition.[23] The changing Arctic thus serves as both a justification for economic expansion and a rationale for multilateral governance structures, such as the Polar Code within the International Maritime Organization and broader Arctic Council negotiations.[24]

Historical
Source: Envisat and CryoSat satellite images from 1950 to 2015 for October
[×10^6 m²]
- 1950:
- 1980: 9218
- 1985: 8562
- 1990: 8409
- 1995: 7650
- 2000: 8458
- 2005: 6996
- 2006: 7693
- 2007: 7345
- 2008: 6997
- 2009: 6649
- 2010: 6913
- 2011: 6084
- 2012: 5657
- 2013: 7427
- 2014: 6926
- 2015: 6756

Multi-Year Sea Ice
Source: National Snow and Ice Data Center (NSIDC) and Arctic and Antarctic Research Institute (AARI)
- Fast ice
- Nilas (0–10 cm)
- Young ice (10–30 cm)
- First zear sea ice (30–200 cm)
- Old sea ice

Sovereignty Coastal States
Source: Legal framework for all marine and maritime activities: United Nations Convention on the Law of the Sea (UNCLOS), 1982

Border Claims
Source: IBRU Centre for Borders Research, Durham University, December 22, 2022
- 2006: Norway
- 2006: Russia
- 2009: Denmark
- 2013: Canada
- 2018: United States

Areas of Ecological Importance
Source: CAFF's Arctic Biodiversity Data Service (ABDS): Seafloor Geomorphology
- Glacial troughs
- Ecological collection points

Temporal

Arctic Tern Conservation Areas and Overlapping Zones of Resource Extraction. This composite map visualizes the spatial entanglement between ecological vulnerability and industrial ambition across the Arctic. The migration routes and breeding grounds of the Arctic tern are superimposed onto zones of oil and gas interest delineated by the United States Geological Survey (USGS). These "assessment units," developed for resource appraisal, highlight regions with high potential for hydrocarbon extraction.

Breeding Areas
Source: CAFF: Arctic Flora and Fauna

Migration Patterns
Source: Ocean Biodiversity Information System (OBIS): Seamap
- June to August: Arctic
- September: Atlantic travel
- November: Arrive in Antarctica
- December to April: Antarctica
- May: Atlantic travel
- June: Arrive in the Arctic

Vegetation Type
Source: CAFF; Classification following CAVM subzones
- Wetland
- Tundra
- Bedrock

- Treeline
Source: Arctic definitions, AMAP

Undiscovered Oil Reserves
Source: USGS, assessment units in the circum-arctic recourse appraisal
[billion barrels]
- Not Assessed
- < 1
- 1–10
- > 10

Arctic Basins
Source: USGS

Oil Infrastructure
Source: Arctic Portal, Nordregio, NSIDC, and USGS
- 2010: Gas terminals
- 2020: Existing infrastructure: Offshore storage, mines, pipelines, power and processing plants, well and platform pads, stations, refineries

Constructing Cartographies

As sea ice retreats to historic lows, the Earth's cryosphere is becoming a focal point for both environmental science and resource exploitation. In this context, specialized infrastructures, such as research stations, satellite networks, weather stations, subsea sensors and installations, icebreakers, and resource exploitation facilities, play a crucial role in shaping our understanding of the region's rapidly changing dynamics. These systems not only monitor environmental change but also facilitate access to previously unreachable territories, enabling the expansion of shipping corridors and economic activity. Far from being neutral, these infrastructures serve as instruments of power, embedding political and territorial claims into the Arctic's fragile and rapidly transforming environment.[25]

Micro-Jurisdictions: Ports
Harbor size according to World Port Index (WPI) classification based on area, facilities, and wharf space
  Large to small

Micro-Jurisdictions: Airports
Airport size based on data from Openflights and Nordregio
  Medium to small

Fiber-Optic Cables
Source: Telegeography and Arctic Cable Company
— 1994–2020: current network
— From 2020: planned network

Shipping Routes
Source: Arctic Portal
— Northwest Passage:
  Ice free from September 2018
— Northern Sea Route:
  Ice free from September 2019
— Transpolar Sea Route:
  Predicted for September 2030

Type of Settlement by Population
Based on Nordregio's classification system
[Inhabitants]
  2,000–5,000
  5,000–10,000
  10,000–50,000
  > 50,000

Road Accessibility Settlements
Categories according to GRIP Global Road Dataset
— Primary
— Secondary
— Tertiary

Railways
Source: Global Railways (WFP SDI-T, Logistics Database)

Temporal

Monsters and Ghosts of the Far North searches for an alternative cartography through which we can rethink relationships across species in the Arctic region, and beyond. This has inspired a pursuit for a medium and methodology that would capture the dynamic nature of these landscapes. As part of this speculative research and design exercise, a digital environment is employed as a testing ground for alternative governance structures and forms of representation, in which we do not radically separate humans from nonhumans. It further questions and searches for ways of disrupting the division of our world and the meaning of current rigid borderlines within distorted, often Eurocentric cartographic projections.

Temporal

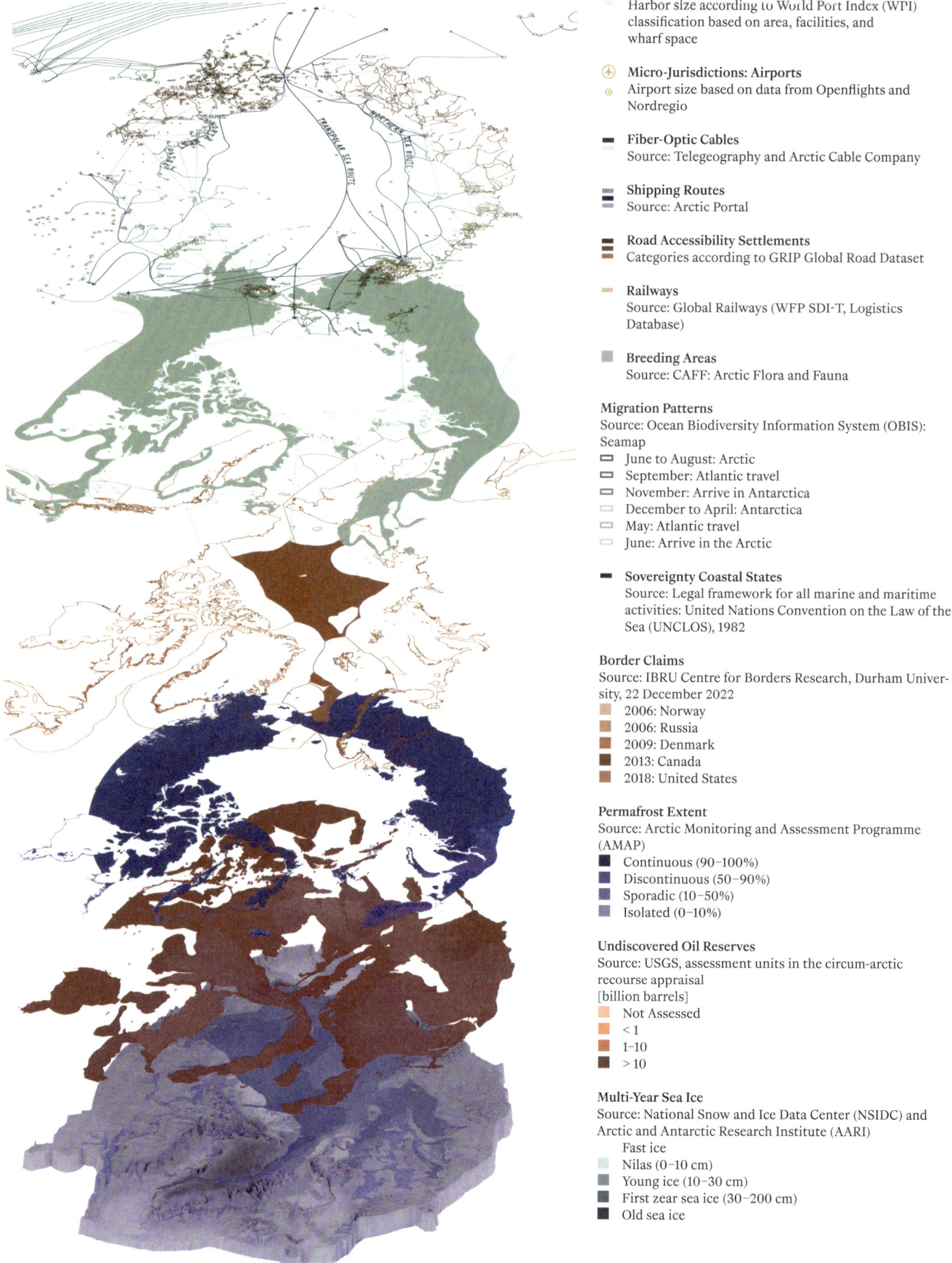

Micro-Jurisdictions: Ports
Harbor size according to World Port Index (WPI) classification based on area, facilities, and wharf space

Micro-Jurisdictions: Airports
Airport size based on data from Openflights and Nordregio

Fiber-Optic Cables
Source: Telegeography and Arctic Cable Company

Shipping Routes
Source: Arctic Portal

Road Accessibility Settlements
Categories according to GRIP Global Road Dataset

Railways
Source: Global Railways (WFP SDI-T, Logistics Database)

Breeding Areas
Source: CAFF: Arctic Flora and Fauna

Migration Patterns
Source: Ocean Biodiversity Information System (OBIS): Seamap
- June to August: Arctic
- September: Atlantic travel
- November: Arrive in Antarctica
- December to April: Antarctica
- May: Atlantic travel
- June: Arrive in the Arctic

Sovereignty Coastal States
Source: Legal framework for all marine and maritime activities: United Nations Convention on the Law of the Sea (UNCLOS), 1982

Border Claims
Source: IBRU Centre for Borders Research, Durham University, 22 December 2022
- 2006: Norway
- 2006: Russia
- 2009: Denmark
- 2013: Canada
- 2018: United States

Permafrost Extent
Source: Arctic Monitoring and Assessment Programme (AMAP)
- Continuous (90–100%)
- Discontinuous (50–90%)
- Sporadic (10–50%)
- Isolated (0–10%)

Undiscovered Oil Reserves
Source: USGS, assessment units in the circum-arctic recourse appraisal
[billion barrels]
- Not Assessed
- < 1
- 1–10
- > 10

Multi-Year Sea Ice
Source: National Snow and Ice Data Center (NSIDC) and Arctic and Antarctic Research Institute (AARI)
- Fast ice
- Nilas (0–10 cm)
- Young ice (10–30 cm)
- First zear sea ice (30–200 cm)
- Old sea ice

As part of this ongoing exercise, we experimented with different configurations of spatial information, each derived from distinct sources, to explore how industrial infrastructures—ports, maritime corridors, sites of resource extraction, research stations—adapt to and instrumentalize the shifting Arctic environment. Map showing the convergence of migratory species conservation areas, permafrost and sea ice extents, as well as resource extraction zones with expanding Arctic sea routes, highlighting the challenges of coexistence where ecological preservation and industrial ambition intersect.

Cryopolitics: Sea Ice Loss and the Construction of Borders

Over the past two decades, territorial surveying projects have supported national claims to the Arctic Ocean floor under the United Nations Convention on the Law of the Sea (UNCLOS). Under UNCLOS, Arctic coastal states can claim an extended continental shelf (ECS) beyond 200 nautical miles if they can provide scientific evidence that the seabed is a natural extension of their land territory.[26] To substantiate such claims, states must collect and analyze scientific data detailing the depth, shape, and geophysical characteristics of the seabed and sub-seafloor. This involves extensive surveys using technologies like multibeam sonar and seismic reflection to map the ocean floor.[27] In this context, authority is no longer asserted through military force but rather through the possession of information, with spatial data emerging as the currency of sovereignty. The sensing revolution has enabled a new spatial ontology that challenges Westphalian notions of sovereignty,[28] where nations have exclusive sovereignty over their territory, by proposing that Arctic territories should be viewed as a three-dimensional volume.[29]

An Arena of Extrastatecraft

In the Arctic, extrastatecraft[30] manifests through infrastructural developments, such as offshore drilling platforms, subsea infrastructures, transport corridors, and specialized monitoring systems, which extend beyond traditional state mechanisms, integrating the region into global markets and geopolitical networks. Simultaneously, environmental features—the dynamics of sea ice and permafrost, the ocean currents, aquatic and terrestrial ecosystems—remain an active participant in the production of space. While Easterling doesn't explicitly categorize natural systems as extrastatecraft, her framework invites us to consider how, in regions like the Arctic, environmental forces interact with infrastructural developments, influencing geopolitical and economic landscapes.

As states and corporations deploy increasingly sophisticated data-driven systems, the Arctic's landscapes are not only observed but actively shaped by the infrastructures that translate environmental phenomena into actionable intelligence, territorial claims, and economic strategies.

The Arctic region thus emerges as a hybrid zone where planetary urbanization, scientific investigation, military strategy, and capitalist extractivism coalesce with environmental dynamics and forge a spatial field in which the built environment/systems of power and natural processes are mutually constitutive. This perspective underscores that environmental governance in the Arctic is not merely about managing nature but about reconfiguring space itself, where industrial and scientific interventions become deeply enmeshed with the evolving character of the natural environment.

Practice-Based Research: Monsters and Ghosts of the Far North

Emerging from a research-driven master's in architecture at the Royal College of Art, this piece explores themes that have informed and inspired the collective project titled Monsters and Ghosts of the Far North, as well as a broader methodological trajectory of research and experimentation.

The concept of a "living map"[31]—ever-evolving and never complete—serves as a powerful metaphor for the ongoing exchanges and transformations of matter. This notion has also shaped the project's methodological approach. Within this framework, "monsters" and "ghosts," as defined by Anna Tsing and her co-authors in *Arts of Living on a Damaged Planet* (2017) become the main agents of an interactive, more-than-human cartography. Monsters mark the entangled, multispecies aftermaths of environmental disruption, while ghosts embody the lingering traces of loss, extinction, and colonial violence—presences that continue to haunt landscapes and systems long after the fact.[32]

One of the main strands of research behind this project was an inquiry into modes of multi-species cohabitation and negotiation of space in light of the ongoing changes in the region. As an interactive multiplayer experience, the ability to digitally embody six different nonhuman characters sought to challenge our current anthropocentric perception and urge to be in control of our environment. In doing so, the multiplayer interaction employs publicly available data from environmental models and builds on monitoring and assessment data from Arctic Biodiversity Data Services,[33] which have the potential to visualize a narrative of global entanglements.

The aim of this prototype was to challenge and subvert existing means of asserting and maintaining sovereignty and geopolitical power by embedding various research elements and outcomes into the aesthetic and mechanics of the digital environment, in which humans are only present by proxy. Our uninterrupted, yet indirect presence underlines our convoluted entanglement. Monsters and Ghosts of the Far North is thus both a platform for and a digital manifestation of "bodies tumbled into bodies."[34]

Monsters and Ghosts of the Far North is a collaborative project by Andra Pop-Jurj and Lena Geerts Danau, realized with the support of the Driving the Human initiative between 2019 and 2021.

1. Philip E. Steinberg, Jeremy Tasch, and Corey Johnson, *Contesting the Arctic: Politics and Imaginaries in the Circumpolar North* (London: I. B. Tauris, 2015), p. 17.

2. The Arctic Council's Arctic Monitoring and Assessment Programme Working Group Boundary (AMAP) incorporates elements of the Arctic Circle, political boundaries, vegetation boundaries, permafrost limits, and major oceanographic features. The region covered by AMAP is, therefore, essentially the terrestrial and marine areas north of the Arctic Circle (66°32'N), and north of 62°N in Asia and 60°N in North America, modified to include the marine areas north of the Aleutian Chain, Hudson Bay, and parts of the North Atlantic Ocean including the Labrador Sea.

3. Susan Schuppli, *Material Witness: Media Forensics Evidence* (Cambridge, MA: MIT Press, 2020), p. 283.

4. C. R. Stokes, J. C. Ely, C. D. Clark, and E. C. King, "A New Record of the Dimensions, Dynamics, and Ice-Marginal Land Systems of a Palaeo-Ice Stream in the Labrador Sector of the Laurentide Ice Sheet," *Journal of Geophysical Research: Earth Surface* 127, no. 2 (2021), e2021JF006123, https://doi.org/10.1029/2021JF006123.

5. IPCC (2019), Special Report on the Ocean and Cryosphere in a Changing Climate. The Arctic is the largest climate-sensitive carbon pool on Earth, storing nearly twice as much carbon in its permafrost as is currently held in the atmosphere. Widespread thawing of near-surface Arctic permafrost is projected this century with very high confidence, leading to a 2–66 percent reduction in permafrost area under a low-emissions scenario (RCP2.6) and a 30–99 percent reduction under a high-emissions scenario (RCP8.5), where RCP stands for Representative Concentration Pathway (scenarios used in climate modeling to explore potential future greenhouse gas concentrations and their impact). This thaw could release up to 240 gigatons of carbon (Gt C) as CO_2 and methane, significantly amplifying global warming. While methane emissions will be smaller in volume (0.01–0.06 Gt CH_4 per year), they could contribute 40–70 percent of the additional warming due to their high radiative forcing potential. See https://www.ipcc.ch/srocc/.

6. Boris Revich and Marina A. Podolnaya, "Thawing of Permafrost May Disturb Historic Cattle Burial Grounds in East Siberia," *Global Health Action* 4, no. 1 (2011): 8482, https://doi.org/10.3402/gha.v4i0.8482. Climate warming in the Arctic may increase the risk of disease outbreaks by expanding the areas where disease-carrying animals can live and survive the winter. Permafrost monitoring since 1970 shows surface layers have warmed by 2–4 degrees Celsius, with a further 3 degrees Celsius increase expected. Between 1897 and 1925, anthrax outbreaks killed 1.5 million deer, and more than 29,000 settlements in the Russian North—especially in Yakutia—are located near burial sites of infected animals. While it is still unclear if warming will release viable anthrax spores, careful monitoring is recommended.

7. Moritz Langer, Thomas Schneider von Deimling, Sebastian Westermann, et al., "Thawing Permafrost Poses Environmental Threat to Thousands of Sites with Legacy Industrial Contamination," *Nature Communications* 14 (2023): 1721, https://doi.org/10.1038/s41467-023-37276-4.

8   Marco Ferrari, Elisa Pasqual, and Andrea Bagnato, *A Moving Border: Alpine Cartographies of Climate Change* (New York: Columbia Books on Architecture and the City, 2018), p. 20.

9   Tom Parfitt, "Russia Plants Flag on North Pole Seabed," *The Guardian,* August 2, 2007. On August 2, 2007, as part of the Arktika 2007 expedition, Russian explorers descended over 4,000 meters beneath the Arctic Ocean to plant a titanium Russian flag on the seabed at the North Pole. This symbolic act aimed to support Russia's claim that the Lomonosov Ridge is an extension of its continental shelf, potentially granting rights to vast underwater oil and gas reserves. The move drew criticism from other Arctic nations, with Canada's foreign minister likening it to outdated colonial tactics.

10 Matthias Heymann, "Search of Control: Arctic Weather Stations in the Early Cold War," in *Exploring Greenland,* ed. R. Doel, K. Harper, and M. Heymann (New York: Palgrave Macmillan, 2016), p. 3.

11 J. Bruun and P. Steinberg, "Placing Territory on Ice: Militarisation, Measurement, and Murder in the High Arctic," in *Territory beyond Terra,* ed. K. A. Peters, P. E. Steinberg, and E. Stratford (Lanham, MD: Rowman & Littlefield International, 2018), pp. 147–65.

12 Marlene Laruelle, "The Three Waves of Arctic Urbanisation: Drivers, Evolutions, Prospects," *Polar Record* 55, no. 1, (2019): 1–12, doi:10.1017/S0032247419000081.

13 "10 Fascinating Historic Maps of the Arctic." *Canadian Geographic,* n.d. Gerardus Mercator's 1595 map was among the first to depict an all-water route across the top of North America and imagined the North Pole as a magnetic rock surrounded by four islands. This speculative geography reflected early European desires to render the Arctic legible—and navigable—for imperial aims. See https://canadiangeographic.ca/articles/10-fascinating-historic-maps-of-the-arctic/.

14 Peter Hemmersam, *Making the Arctic City: The History and Future of Urbanism in the Circumpolar North* (London: Bloomsbury Visual Arts, 2021), p. 158. Danish architect Alfred J. Råvad, influenced by the City Beautiful movement, proposed transforming Greenland's scattered, isolated settlements into a modern capital. His plan featured clearly defined sectors—governmental, commercial, residential, and industrial—with grand boulevards, public parks, and civic monuments. Råvad's concept extended beyond infrastructure; it aimed to "nationalize" Greenland by settling Danish and Icelandic families, thereby asserting Danish cultural identity and countering perceived German influence. This early vision contrasts sharply with later developments: following Hitler's 1940 invasion of Denmark, the establishment of permanent US military bases redefined Greenland's strategic role in the Arctic.

15 Hemmersam, *Making the Arctic City,* p. 164.

16 "Stereographic," ArcGIS Pro Documentation, Esri, https://pro.arcgis.com/en/pro-app/latest/help/mapping/properties/stereographic.htm.

17 For a recent geological and geopolitical analysis of the Lomonosov Ridge and its role in Arctic territorial claims, see Marc C. Pigott et al., "The Lomonosov Ridge, Central Arctic Ocean: The World's Largest Sliver of Continental Crust," *Journal of the Geological Society* 182, no. 2 (2024), https://pubs.geoscienceworld.org/gsl/jgs/article/182/2/jgs2024-095/649821/The-Lomonosov-Ridge-central-Arctic-Ocean-the-world.

18 For instance, the integration of multi-core CPUs, GPUs, and FPGAs in satellite systems enables complex tasks such as hyperspectral image classification and deep learning inference to be performed onboard, facilitating timely decision-making in applications ranging from climate monitoring to disaster response.

19 European Space Agency, "Permafrost Monitoring from Space: A Review," ESA Climate Office, n.d. Despite advances in remote sensing, permafrost carbon emissions remain insufficiently integrated into global carbon budget models, according to the IPCC. See https://climate.esa.int/en/news-events/permafrost-monitoring-from-space-a-review/.

20 Terrain, Arctic DEM, https://livingatlas2.arcgis.com/arcticdemexplorer/.

21 The region covered by AMAP is, therefore, essentially the terrestrial and marine areas north of the Arctic Circle (66°32'N), and north of 62°N in Asia and 60°N in North America, modified to include the marine areas north of the Aleutian Chain, Hudson Bay, and parts of the North Atlantic Ocean including the Labrador Sea. AMAP was launched in 1991 in response to growing environmental concerns in the Arctic, establishing a science-based approach to monitor, assess, and guide policy and safeguard Arctic ecosystems.

22 E. A. G. Schuur et al., "Vulnerability of Permafrost Carbon to Climate Change: Implications for the Global Carbon Cycle," *BioScience* 58, no. 8 (2008): 701–14, https://doi.org/10.1641/B580807. In oxygen-depleted conditions within the seasonally thawed active layer, methanogenic archaea reduce hydrogen and acetate to methane, while aerobic methane-oxidizing bacteria can consume up to 90 percent of that methane before it escapes; however, warming-induced permafrost thaw risks liberating vast stores of preserved soil organic carbon, potentially accelerating microbial methane emissions and climate feedback.

23 D. L. Gautier, K. J. Bird, R. R. Charpentier, A. Grantz, D. W. Houseknecht, T. R. Klett, T. E. Moore, J. K. Pitman, C. J. Schenk, J. H. Schuenemeyer, K. Sørensen, K., M. E. Tennyson, Z. C. Valin, and C. J. Wandrey, "Assessment of Undiscovered Oil and Gas in the Arctic," *Science* 324, no. 5931 (2017): 1175–79.

24 K. Keil and S. Knecht, eds., *Governing Arctic Change: Global Perspectives* (London: Palgrave Macmillan, 2017).

25 Steinberg, Tasch, and Johnson, *Contesting the Arctic,* p. 63.

26 Ragnhild Groenning, "Exploring Continental Shelf Claims in the Arctic – Infographic," The Arctic Institute, https://www.thearcticinstitute.org/wp-content/uploads/2017/06/TAI-Infographic-ContinentalShelfClaims.pdf.

27 National Oceanic and Atmospheric Administration (NOAA), Extended Continental Shelf Project, NOAA Ocean Exploration, https://oceanexplorer.noaa.gov/about/what-we-do/media/ecs-project.pdf.

28 Stephen D. Krasner, *Sovereignty: Organized Hypocrisy* (Princeton: Princeton University Press, 1999), pp. 9–25. Westphalian sovereignty refers to the principle that each state has exclusive authority over its territory and domestic affairs, free from external interference. This concept emerged from the Peace of Westphalia in 1648, which concluded the Thirty Years' War and established a new political order in Europe based on the coexistence of sovereign states. The principle emphasizes territorial integrity, non-intervention, and legal equality among states, forming the foundation of the modern international system. However, Krasner argues that the application of these principles has been inconsistent, leading to what he terms "organized hypocrisy" in international relations. He delineates four distinct attributes of sovereignty: international legal sovereignty (recognition by other states), Westphalian sovereignty (non-intervention), domestic sovereignty (effective internal control), and interdependence sovereignty (control over trans-border flows).

29 Corine Wood-Donnelly, "Sovereignty Cubed: The Arctic as a Territorial and Ontological Volume," *European Journal of Social Theory* 28, no. 2 (2024), doi:10.1177/13684310241270533.

30 Keller Easterling's concept of "extrastatecraft" refers to the ways in which infrastructure operates beyond traditional state mechanisms to exert power and influence. It encompasses the networks, systems, and protocols, such as free trade zones, communication grids, and global standards, which shape urban and economic landscapes without direct governmental oversight. These infrastructural elements can create new forms of governance and control, often faster and more pervasively than official state policies. In the context of the Arctic, extrastatecraft can be observed in the strategic development of infrastructure by various actors, including corporations and international consortia, aiming to exploit the region's resources and navigational routes. As the Arctic becomes more accessible due to climate change, these non-state entities may establish operational frameworks, such as shipping lanes, energy extraction sites, and communication networks, which function with minimal state intervention, effectively reshaping the geopolitical dynamics of the region through infrastructural means.

31 Frédérique Aït-Touati, Alexandra Arènes, and Axelle Grégoire, *Terra Forma: A Book of Speculative Maps* (Cambridge, MA: MIT Press, 2022), p. 19.

32 Anna Lowenhaupt Tsing, Nils Bubandt, Elaine Gan, and Heather Anne Swanson, eds., *Arts of Living on a Damaged Planet* (Minneapolis: University of Minnesota Press, 2017).

33 See the Conservation of Arctic Flora and Fauna website, https://abds.is/.

34 Tsing et al., *Arts of Living on a Damaged Planet,* p. 13.

Teresa Fankhäncl and Max Hallinan

In a Manner of Speaking: From Bits to Bricks

Fig. 1: The program alerts a user to design problems. URBAN 5, Architecture Machine Group, 1967.

Modeling a Model of Things

The first words spoken by a computer to an architect contained a warning. "Ted, many conflicts are occurring" (fig. 1). Ted, the unknown user in the Architecture Machine Group's experiments, was talking to URBAN 5, an experimental program designed to augment human architects. As part of a two-decades-long research project exploring computational design and human-machine interaction, MIT's researchers incorporated a language generator into the fifth iteration of their URBAN series of computer programs. The model could verbalize conflicts in the designed structure in an ostensibly chummy way. The machine was supposed to be a friend, a helper, a learner, and a partner to the human.

Several decades later, in 1994, Nicholas Negroponte, one of the founders of the Architecture Machine group, described the translation of a "model of things" to the individual recipient, from inaccessible knowledge stored in machine code to something a human could process: "There are no words and there are no animations or anything that we would think of as a medium in this model. You could transcode those bits from the form they arrived in, a model, into a variety of media. Bits are bits."[1] He goes on to clarify that the way such a model might choose to reveal itself should be personal as it has to meet its users where they are: "Whatever transcodes it will be doing it not only for you, but presumably knowing you. So let's say it's gonna speak it. It would speak it differently to you than to me than to someone else. It would obviously focus on different areas. If I were a fisherman it would concentrate more on the maritime weather, and the high tide and low tide and the size of the waves, which when I'm sitting in my office in Cambridge, Massachusetts, is of absolutely no consequence."[2] In such a "model of things" knowledge itself, the accumulation of data in bits and bytes, is without meaning if it cannot be applied to a given context; and the weather report's concrete language, the medium the model is translated into, the syntax and lexicon used, is key to suss out meaning. Today Negroponte's model of things could very well refer to generative AI, a term he himself helped establish in architecture beginning in the 1960s.

Build, Don't Talk?

The team behind URBAN 5 must have been aware of architecture's longstanding love-hate relationship with language, an inherent uneasiness with words as an integral part of architecture. In his groundbreaking study of the language of modernism architectural historian Adrian Forty parades a roster of well-known architectural theorists who argue for or against the inclusion of the so-called "architectus verborum," the architect of words, in the ranks of architectural professionals.[3] Part of the animosity toward verbalizing architectural thought is the long-standing assumption that talking about an object or building is incompatible with the thing itself. No words can express fully what the senses experience. To talk about a building is to destroy what makes it unique.[4] "Build, don't talk," Ludwig Mies van der Rohe said tersely.[5] Forty's book, therefore, is less dictionary and more encyclopedia, a critical vocabulary of "encounters with things" but also more importantly of encounters with what they are not.[6] Words help to draw out meanings and differences, sharpen societal concepts and aid in defining the purpose of design within the contentious discourses of "good" design.

In 1993, right as the World Wide Web became accessible and as software was making its way into architecture schools, Stanley Tigerman contributed to the first issue of *ANY* magazine a provocative iteration of this age-old question of whether to speak or not to speak about architecture. Would language ever be able to have any effect on architecture, he asked? "Can the penetration of language sufficiently contaminate a discipline so as to signal its imminent overturning, or will architecture remain as dumbly impenetrable to change as earlier extrinsic attempts to influence it have shown it to be?"[7] Tigerman was not interested in AI but, rather, in an argument against the possibility of unifying language with architecture. "For a parasite (writing)," he wrote, "to contaminate a host (architecture) by being IN it, to get under its skin, the host must have cells both deferential *and* vulnerable to those of the parasite."[8] Now, how deferential and vulnerable is architecture to language? And has language really been extrinsic to architecture, OUTSIDE rather than IN it?

Temporal

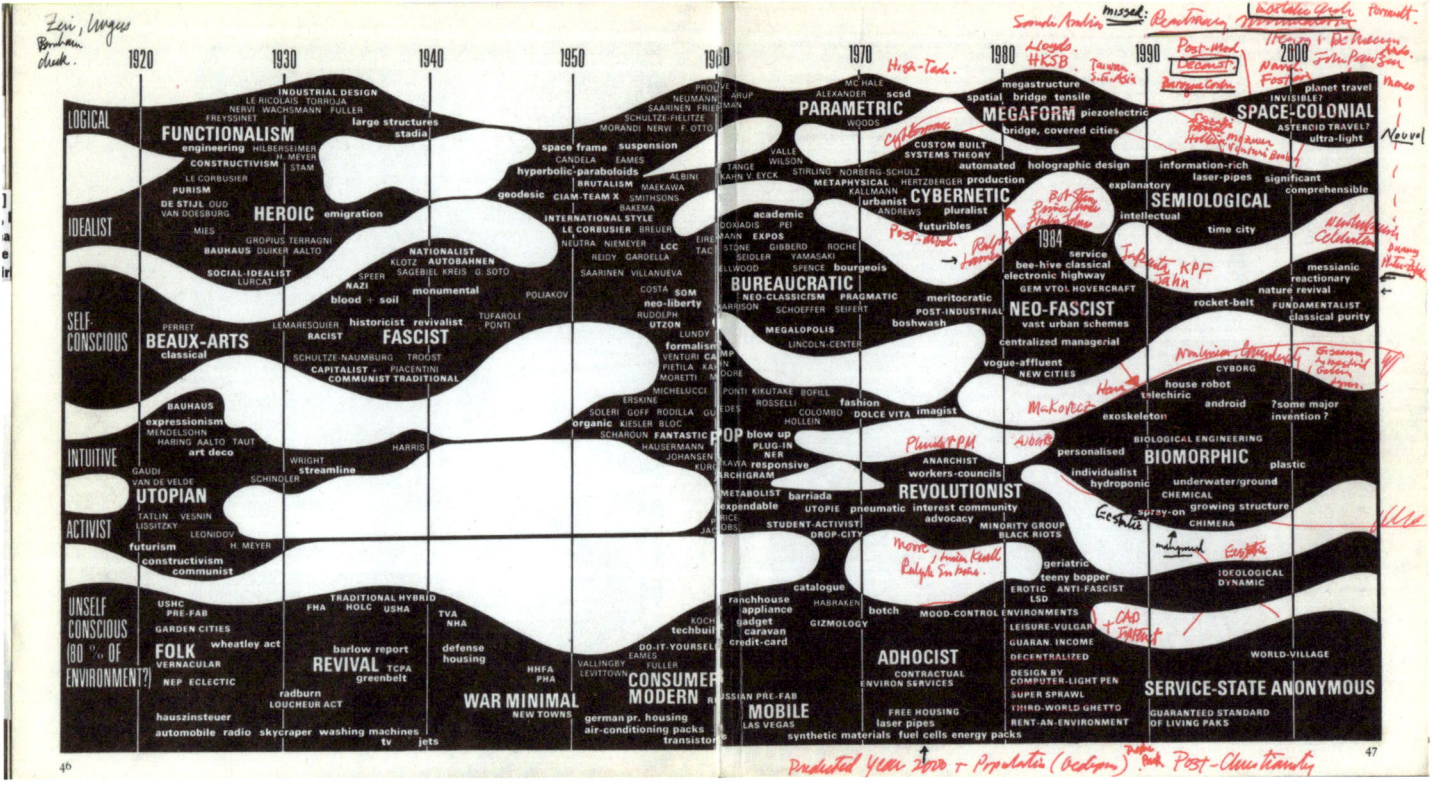

Fig. 2: *Evolutionary Tree to the Year 2000*, Charles Jencks, 1969. With revisions and notes, ca. 1999.

There are ubiquitous metaphors that seek to establish some kind of kindred spirit between architecture and language—references to architecture as grammar or syntax—that have pervaded architectural thinking such as Christopher Alexander's aptly named *Pattern Language* or the more thesauric approach of Charles Jencks's *Language of Postmodern Architecture*. Language here is not just a medium of instruction and communication; it is a code. A number of modern architects have been vocal about their dislike for rendering, perhaps most famously Adolf Loos who, in 1924, declared he had no need for it because

"[g]ood architecture, how something is to be built, can be written. One can write the Parthenon." Loos's example highlights how an architectural practice that depends exclusively on the execution of established rules (in ancient Greece, at a time when there were no architects in the modern sense of the word) could be written out or encoded in language. Such a practice of describing a structure procedurally could also be called programming. If we can call the Parthenon's written architecture expounded by Loos algorithmic, Charles Jencks's work, by contrast, could be considered as the equivalent of an ever-shifting database. His continuous revisions of his *Evolutionary Tree of the Year 2000* (fig. 2) were an iterative labeling process of a vector field that started to detach from the linear relations of traditional repositories of architectural knowledge such as dictionaries or databases. And while the structures subsumed in his evolutionary construct were part of an academic discourse surrounding the canon of architecture, even if they took the idea of a contemporaneous canon ad absurdum, his *Daydream Houses of Los Angeles* achieved something altogether more riveting. Driving around the suburbs of LA Jencks recorded chance encounters while house-chasing local vernacular architecture that "bastardized" traditional styles such as ranch houses or haciendas. In a tongue-in-cheek name game, he churned out labels for the kitschy and the excessive he encountered based on iterations of historical references (fig. 3).

Daydreaming with AI

New technologies, especially generative artificial intelligences, revive the age-old discussion about text-text, image-text, and object-text pairs in architecture. They now enable a bi-directional, seemingly automatic, conversion of written descriptions into images and found images to texts. With large language models (LLMs) computers have learned to mimic

Fig. 3: *Daydream Houses of Los Angeles*, Charles Jencks (New York: Rizzoli, 1978).

human speech beyond URBAN 5's limited preprogrammed responses. And with advances in computer vision, neural networks—approximations of the inner workings of firing neurons inside a human brain—have learned to discern between representations of real-world things in images. Unlike in humans such learning only happens with the help of preexisting vast collections of processable data, be it in the form of textual sources, images, or videos that allow an extrapolation of similarities and conventions. Without the millions and now billions of datasets collected by services like WordNet, ImageNet, Common Crawl, or LAION, computers would toddle on. That the makeup and interpretation of these datasets is not without pitfalls is by now widely recognized through these systems' built-in biases and a propensity to hallucinate, that is, to invent information they do not contain without making such fabulations apparent. Some earlier image generators such as DeepDream (many of the early neural networks consciously leaned into a somnambulist language, from Deep Daze to the Raymond Chandleresque Big Sleep) tended to limit their creations to clearly definable groups of images that exposed a system's underlying database as much as the way it was parsed.[10] If the data's classification contained mostly pictures of dogs, dogs is what the generated images skewed toward (fig. 4).

This means that there are no Platonic ideals in AI, no models of eternal truths or forms to refer back to. Rather, AI's "model of things" is a dynamic vector space that plots adjacencies and calculates probabilities based on what it has ingested. As Mario Carpo put it, every raw, unstructured, and unlabeled collection of images scraped from the internet and used for training is a potential canon of architecture.[11] Mapping one system of knowledge onto the other, jointly training to align a text encoder and an image encoder, for example using Contrastive Language-Image Pre-training (CLIP), is a relatively new method. Much of this learning happens as self-supervised learning, which is, if at all, only grounded in a small sample set that contains a so-called ground truth, that is, information and meaning verified by humans. Everything else is inferred by continuous adversarial comparison. Like a kid on a long car ride, one part of the network keeps asking "are we there yet" and the other keeps responding "no" until the destination, a final image-text pair, is reached.

A recent quantitative study of such natural language encoding in Midjourney, arguably the text-to-image platform most widely used by architects, tries to pry open the black box of such a joint learning process.[12] Leaning heavily on the idea of "style" as an already encoded knowledge system in architecture with defined design "rules," the researchers asked ChatGPT, a text-to-text generator, for a description of a generic Art Deco interior: "Geometric shapes bold colors luxurious materials streamlined forms symmetrical patterns decorative motifs plush interiors lavish embellishments glamorous aesthetics modernist design." Whereas modernist writing had preferred abstractions and generalizations, including a limited number

Fig. 5: "Art Deco interior" created using Midjourney. Joern Ploennigs and Markus Berger, ca. 2024.

of nouns, discarding adverbs, attributes or metaphors, the new architectural language of some AI generators seems to thrive on the use of experiential qualities of spaces and an overabundance of descriptive imagery. In a second step, after they had prompted Midjourney to create the image (fig. 5), the study's principal investigators asked the same program for an image-to-text conversion back that revealed how deeply the idea of "style" is encoded in the makeup of generative AI such as Midjourney. All of Midjourney's generated descriptions featured the key phrases "Art Deco" and "geometric patterns" rather than ChatGPT's more descriptive keywords used as input by the researchers.

Prompting and Probabilities

Common to all such efforts in generating content is that machine learning atomizes preexisting systems of meaning curated by humans: images are successively disintegrated until they contain nothing but "noise," and words are stripped down to their stems. Out of the protean soup of information arises a learned language based on stochastic principles that generates sentences according to the most likely words following one another and creates images by comparing them to the median standard established by the learning process. Linear categories of meaning that are established for traditional canons—from dictionaries and encyclopedias to database structures of archives and collections—become vectorized, taken out of the traditional structure that can be represented in relational SQL databases, and mapped onto a decision tree. Language, in this new system, may seem like a natural language because, much like URBAN 5's chatbot, it follows human rules of speech in the interfaces we access. Yet, such appearances are deceptive.

LLMs and other generative AI possess none of the formal or conceptual structures that humans use to interpret data, and they never will. Comprehension, in this sense, is not the goal of deep learning. An LLM understands its training data simply as a long sequence of bytes in which many subsequences follow patterns. These patterns certainly correspond to structures that are meaningful to humans. Purposeful arrangement of characters or pixels is part of the way that humans make verbal or visual phenomena particularly significant. To express love, for example, we might select and arrange characters from the English alphabet to follow the culturally significant pattern "I love you." But the LLM does not model that significance. It merely knows the pattern as a more or less common arrangement of bytes in its training data. The ability to respond coherently to a complex prompt is not any grasp of the prompt's meaning but simply an accurate prediction that one pattern of bytes follows another. "I love you," for example, might be the most probable end to a sequence that begins with the characters "Do you love me?" And if these phrases don't occur in the training data, then the LLM will not have this language for love. The model knows only what it has been given to know and knows it purely in the form of probable relationships between sequences of bytes.

"[A] programming language represents a 'modeled world' that, in the case of general-purpose languages, has the same universal character as the underlying hardware, but structures access to that full space of computation in specific ways," writes researcher Bernhard Rieder in his study of programming languages as "engines of order."[13] Natural language, as usurped by AI, equally structures access to the data it has ingested. But while human-made programming languages are translated in a direct way into machine code and have a predictable outcome, this is not so with generative AI. What you see (as words) isn't entirely what you will get (as a one-to-one reversible response to your prompt). Still, language, to use Stanley Tigerman's metaphor, has crawled under the skin of architecture by way of the AI used by architects. In fields such as architectural history, with its centuries-old discourses on historiography, taxonomy, and a more recent grappling with the underlying biases in existing knowledge systems, AI's shorthand understanding of language often remains stuck in the coffee-table-book version of buildings as out-of-the-box styles. It cannot independently discover or connect to any ground truths that humans both continue to establish and knock down.

Predictability and patterns are what it all boils down to, both for past and present architecture. In the service of architecture a generative AI could be expected to behave like one of Frank Lloyd Wright's draftspersons, who weren't told exactly what to draw down to the last detail. This level of precision would have amounted to Wright doing the drawing himself. Instead, Wright directed the drawing at a high level and depended on the draftspersons to fill in the blanks as he would have done. Because they were trained on the "dataset" of Wright's previous work and feedback from him in the studio, they knew the patterns that comprised his style and could make good guesses about which pattern should be used to complete the drawing at hand. The usefulness of the architectural draftsman of yesterday, and perhaps the architectural generative AI of tomorrow, rests on this ability to fill in the blanks, to complete the sequence of bytes, according to patterns found in context provided by the architect. Getting the desired result from these assistants, whether human or computer, is more about the curation of training data than making precise requests and then fine-tuning the result through rounds of feedback.

What we should not expect from these models is anything resembling the ingenuity of Frank Lloyd Wright himself. The entire aim of these models is to produce output that is consistent with patterns found in the training dataset. If we could attribute any sort of preference to these models, it would be a preference for doing things the way they are most often done, an essentially conservative preference for the conventions and traditions established by the training data. Wright was certainly fluent in the architectural traditions of his time, but he used that history, arranged those bytes, exactly in ways that were not predicted. These models, being prediction machines, have no way of producing the improbable. They could be, and sometimes are, directed to use a higher degree of randomness in their predictions. But this is a poor approximation of Wright's vision for latent patterns just at the edge of human architectural experience. These latent realities are no more random than they are predictable. To perceive the latent pattern, one must be in direct contact with the underlying human experience, a connection these models do not have. For this we still need humans. And it will be an interesting challenge for those humans to get their model to render this new pattern that is inconsistent with and unpredicted by its training data.

In the introduction to his "Words and Buildings" Adrian Forty pointed out how any inquiry into the language of architecture inevitably has to return to the core of the practice of architecture. The search for the perfect representation of a building has haunted architecture for centuries. The qubits of architecture, the smallest possible carriers of architectural information, remain a subject of great debate. Is it the initial sketch? Is it an orthographic drawing? Is it only the built structure? Is it an algorithm? Or is it, perhaps, the string of words in a prompt or the structure of a database? Such questions go to the heart of a cultural practice that, over the last 200 years, has been in near-constant technological upheaval: What exactly is architecture and what is it that architects do? With the advent of artificially intelligent generators, language and the use of large-scale databases have received renewed potential. If so, studying the underlying structure of generative AI and its relationship with the data it represents will become a new challenge facing not just practitioners but historians and scholars alike. Or, in the words of our chummy digital partner: "Ted, how long are you going to postpone resolving this conflict?"

1 Ted Nelson, "The Human Side of the MIT Media Lab," February 23, 1994, p. 3, https://archive.org/details/Ted_Nelson_The_Human_Side_of_the_MIT_Media_Lab_1994-02-23/page/n1/mode/2up?view=theater.

2 Ibid.

3 Adrian Forty, *Words and Buildings: A Vocabulary of Modern Architecture* (London: Thames & Hudson, 2004), p. 11.

4 Ibid., p. 12.

5 Ibid., p. 13.

6 Ibid., p. 15.

7 Stanley Tigerman, "Writing IN Architecture," *ANY 0: Writing in Architecture, New York* (May/June 1993), p. 39.

8 Ibid., p. 41.

9 Adolf Loos, "Regarding Economy," in *Raumplan vs. Plan Libre: Adolf Loos and Le Corbusier, 1919–1930*, ed. Max Risselada and Beatriz Colomina (New York: Rizzoli, 1988), p. 139.

10 Kyle Steinfeld, "Machine Hands on Flaws to Machine: The Surprising Sources of Bias in Machine Learning Models," *Architectural Design* 94, no. 3 (May/June 2024): 102–09.

11 Mario Carpo, "Every Dataset is a Canon," *Architectural Design* 94, no. 3 (May/June 2024): 14–19.

12 Jörn Plönnigs and Markus Berger, "Generative AI and the History of Architecture," in *Decoding Cultural Heritage: A Critical Dissection and Taxonomy of Human Creativity through Digital Tools* (New York: Springer Books, 2024), pp. 23–45.

13 Bernhard Rieder, *Engines of Order: A Mechanology of Algorithmic Techniques* (Amsterdam: Amsterdam University Press, 2020), p. 109.

Anna-Maria Meister and Rafael Uriarte

Forgetting as a Feature, Not a Bug:
The Intelligence of Loss in the Archive

From the distinct viewpoints of two authors, this text explores how disappearance—both from archives and within them—intersects with forms of data erasure driven by Artificial Intelligence. In the following paragraphs, Anna-Maria Meister's text is left-aligned and Rafael Uriarte's text is right-aligned, while the jointly written text is justified, and the quotations are centered in italics.

Anna-Maria Meister and Rafael Uriarte: Archives are core sites of destruction, decay, and loss. Whether through man-made destruction, be it intentional, arbitrary, or accidental—such as changes in institutional policy, forgetfulness, deliberate evidence destruction, wars, migration, or even the act of curating an exhibition—or so-called natural disasters like earthquakes, fire or water have eradicated much of humanity's stored matter, and keep doing so. From slowly dripping water under broken roofs or dropped bombs, archives are not "safe." Small-scale destruction and micro damage aided by the long durée of the archival perspective can be just as devastating. Be it undetected or untreated mold, a colony of paper fish, or simply crumbling paper, archives and their matter and data continuously disintegrate under human watch. And, not least, archives are full of things no one can access, rendering things lost not due to their absence but their lack of retrievability within the archive itself: from vast collections hidden away in depot spaces forgotten for decades, or a misplaced model on a shelf, to digital-born or digitized data stored on obsolete hardware, operated with deprecated software, defined with superseded or discontinued standards. Moving toward an age of "peak storage" (a limit not of technical expansion, but of sustainable storage where meaningful retrieval reaches its limits), we ask what can be learned through speculative destruction in the face of "peak storage," mass migration, and the predicted AI energy crisis?

The entropic nature of archives, where the very processes designed to preserve knowledge—whether on paper or servers—inevitably lead to transformation, selective preservation, and *loss*. While in art and architecture history, narratives of loss and decay have started to play a more important role in historiography, AI is hailed as a promise *against loss*. But what if we treated forgetting not as a bug, but a feature? AI's processes of erasure follow distinct patterns that may reveal new ways of managing archival sustainability. Its mechanisms of forgetting and loss—such as data selection, pruning, model drift, algorithmic obsolescence and capacity to (almost) randomize choices—offer alternatives to human-driven/caused archival destruction. Rather than treating AI as an optimized alternative for pattern recognition in archives, this research explores how AI's patterns of forgetting might reshape our understanding of archival decay and persistence. By taking loss and erasure seriously—not as failures, but as integral to knowledge systems—we try to explore how it might be an underappreciated field of exploration, one where its (at least seemingly) irrational and unpredictable human shape might be a form of intelligence AI could learn from.

Written from two perspectives, this essay approaches the intersection of archival disappearance and AI-driven erasure. What does the pattern of loss look like and can it be modeled? Reversely, can humans gain insights from AI-driven loss into archival decay? This essay takes loss not as a problem, but as a promise of other ways of knowing through investigations into what is lost with which intentions, what is memorable only because it was lost, or how forgotten materials incorporate intelligence.

Setting Up the Site:
AI in the Archive

Anna-Maria Meister: When AI is discussed in archival studies (and potentially more so in politics), it is hailed as long-awaited liberation of information from dusty matter. The narrative promises automated metadata generation from estates, findability, effortlessly sorting boxes of materials into neat categories in online databases. In short, AI is not only expected to replace the human labor of the archivist, but also to create knowledge categories and structures for future research. Whether that would be a desirable future is to be discussed; it definitely is not a materialized present. And even if—or when—AI will take on archival tasks as they are imagined by engineers and politicians, we pose that the archive is made not only from known (or even knowable) information, but from a material presence that is heterogeneous, constantly changing, and always incomplete. Hence what this essay suggests is that AI might learn less from being fed archival material, and more from the processes of its disappearance inherently embedded in the archive than from its images, documents, and plans.

Rafael Uriarte: Simply training these systems on the surviving, selected, and often biased records risks overlooking and amplifying these absences, particularly in cases of historical silences or arbitrary omissions, thereby reinforcing the very conditions that produced them.
We argue for a shift toward "learning" that engages with both what is present and what is missing (otherwise, would it be missed?). This involves exploring how AI methodologies can move beyond merely reflecting presence in the outputs of their algorithms, toward integrating an understanding of loss of archival repositories. We will propose strategies such as developing metadata structures that explicitly model absence, designing interfaces that visualize and contextualize gaps, and refining AI models to account for the inherent incompleteness and biases within these records. Ultimately, our aim is to leverage these approaches to foster a more critical, nuanced, and historically conscious engagement with AI that grapples with both presence and absence.

The Living Archives: Ecological, Human, and Digital Entanglements

"The Earth itself is, as we know, a heap of rubble from a past future, and humanity the thrown-together, bickering community of heirs to a numinous yesteryear that needs to be constantly appropriated and recast, rejected and destroyed, ignored and suppressed so that, contrary to popular belief, it is not the future but the past that represents the true field of opportunity. That is precisely why its reinterpretation is one of the first official acts of new governing regimes." [1]

Architectural archives are porous, leaky, and unstable. The sequence of processes prior to things entering an archive range widely: from a request of heirs to empty their (usually) father's office, usually filled with a variety of boxes with project files, plan rolls, and models, often mixed in with personal affairs, calendars, diaries, or letters, to a pre-agreed-upon donation of selected projects from so-called important offices. Much has been written on the (problematically) hermetic nature of archives in the disciplinary sense of stabilizing and protecting canonical views. But where more porosity is necessary on a disciplinary level, archives have long been leaking and therefore porous. Reversely, they always were host to uninvited "guests" of all kinds. Never have they been a stable space with clearly defined objects—despite all rhetorics of sorting according to genre, medium, or discipline. In fact, archivalia themselves constantly undo established categories of the archive. Some entered as single, forlorn objects found in obscure places; others were left between the pages of books containing unlabeled photographs or copies without copyrights or were dismissed as "personal items" such as diaries, letters, or photo albums. All these objects, however, sew not only doubt: they bring with them their very own flora and fauna—an archival reality that archivists and restorers try to prevent by putting new acquisitions into quarantine, but which nevertheless diffuses into the archive eventually. If one takes any archive on any day and zooms in, one would find not a mostly hollow space to be filled with dead matter, but a flurry of paper fish, microbes, spores, and algae living together with mice, humans, and paper.[2] Thus, archives are, quite literally, living entities. Rather than understanding them as a collection of things, this essay therefore rephrases the conditions of loss accordingly:

What does "loss" mean if the archive is a habitat populated by a host of diverse living species? What information is really stored in it if one learns from biome analysis and biological milieu studies? If one takes recent concepts of soil, where diverse matter lives in complex, interdependent relationships while creating something often explained as "matter"—namely, soil—the question of material loss equally becomes an interspecies venture.[3]

Reading the archive this way extends existing framings of the archive as a living organism in and of itself, where it was described as a habitat of dynamic processes of diverging temporalities and circulating materials—and not least the humans handling them. Our approach lays the foundation for an understanding of the processes within the archive not as separated from each other (and neither from an "outside"), but as an ecosystem of objects, information, and processing embedded in a larger context. This essay, however, further proposes a move from the metaphor into the *actual* flora and fauna of the archive. It aims to include animated matter into the fold of the archival ecosystem to attempt to fully assess effects and motivations when speaking about material configurations and, ultimately, loss. Thinking back to soil, also in the archive, humans and other living agents are inseparably tied in the production of matter as well as its destruction. Even more, they are not only tied, but themselves are part of the materials within archives; they form and produce and destroy a unified, entangled, enclosed mass of diverse things and beings of diverse scales; they permanently touch, devour, create, and destroy each other. The neat illusion of categorical separation in the archive is but a temporary scaffolding.

Digital archives, too, are vibrant networks where data, metadata, developer, user, as well as content producer interactions evolve. They experience their own forms of decay, such as data corruption, bit rot, and broken links, while also undergoing renewal through updates, migrations, and reinterpretations. Moreover, new and existing algorithms, with their outputs and unforeseen impacts, actively reshape these landscapes, altering content flows, user experiences, and even introducing biases that echo the unpredictable interferences found in natural ecosystems. In this dynamic environment, AI emerges as a living organism within the digital ecosystem—adapting, learning, and co-evolving with its environment to shape the ever-changing landscape of digital archives.

Digital archives and AI engage in an uneven push-and-pull, a continuous, asynchronous feedback loop through which each influences and reshapes the other. This interaction reflects the rapid, unpredictable evolution of AI processes, underscoring a complex interplay where both domains prompt, challenge, and refine each other. As digital archives expand and transform—with influxes of new data, updates in metadata, shifting user interactions, and inevitable loss—AI systems are regularly retrained and recalibrated. Yet, it is not only the data that is in flux; the algorithms and models themselves—exemplified by the rapid evolution of large language models (LLMs)—are constantly refined. This evolution and the very nature of these models produce nondeterministic outputs that defy preset patterns and lead to emergent reconfigurations within the archive. The iterative process enables AI to adapt to new patterns, while its interventions, such as data flow reorganization, censorship, and bias mitigation actively reshape the digital landscape. These sometimes-serendipitous responses drive the archive's evolution, creating new structures, exposing hidden gaps, and suggesting novel interpretations. In this way, AI works not only as a transformative tool but also as an integral participant within the digital ecosystem by continuously influencing and being influenced by the living landscape of digital archives.

Just like physical archives are not separable from their outsides, digital archives are not silos but interconnected ecosystems that offer complementary insights into heritage. For instance, a physical archive might reveal details such as handwritten marginalia on architectural blueprints, wear patterns on documents, or the spatial arrangement of boxes that hints at historical cataloguing practices. These traces provide depth that often eludes digital capture. In contrast, digital archives organize information through dynamic metadata and evolving content, shaped by algorithmic curation, with ranking, grouping, and displaying materials based on content and user behavior, as well as real-time user interactions, such as queries, searches, and tagging. These interactions reshape the archive's structure, uncovering hidden relationships among records and across collections that no human curator could anticipate. Yet, digital archives, too, reveal accidental traces and discoveries, such as metadata left behind from previous categorizations, broken hyperlinks indicating forgotten schemas, or corrupted data that surface during migration or analysis.

While organizational methods may differ, one relying on materiality and physical context, the other on digital connectivity and computational analysis, each system enriches the other, creating a holistic landscape where the nuances of physical existence inform and are amplified by the expansive, interconnected digital realm. Both physical and digital archives are subject to the pressures of decay and regeneration, and each is enriched by the interplay of human and nonhuman agents. This synthesis underscores the argument that archives are living entities, complex, evolving, and defined by the dynamic balance between presence and absence.

The (Human) Database in the Archive

If a drawer has not been opened in twenty years, do the archival documents stored in it still exist? A familiar question from philosophical discourse, for the archive (both a database or a human observer) one might adopt a phenomenological or idealist point of view: things that exist in space and time only exist *as archivalia* when or while they are locatable in an archive. Whether something is an archival document, then, is a temporary condition rather than an ontological category. And yet, things are not as simple as stated above. Yes, objects in archives are archivalia for the duration of their stay there (or their retrievability). But what about documents produced as records, but lost before being found by others? What about stories told, but never recorded? A letter read but not kept? What differentiates memorabilia from archivalia, and memory from history? Something might become an archival object *when it is treated as such.*

The question of meaningful existence in an archival sense of the word (meaning, existing as accessible material and/or information), is enhanced by the prevalence of what has been called "embodied knowledge" in archives. Archivists tending to the documents in understaffed infrastructures over time become the human databases for their retrieval. Institutionalized archives are built to last centuries, while the humans working within them stay years, maximally decades. Eventually the people who have internalized the location of things, the shape of the processes of handling them, the shortest way to a specific document and the forms one needs to navigate to receive, file, and show them, will leave, taking what we might call "embodied database knowledge" with them. A different kind of knowledge, difficult to externalize and even more difficult to categorize, and yet it often provides the most effective and complete access to the archive's holdings.

The permanent externalization of anecdotes, conversations, encounters, emails, and, not least, things observed in the handling and sorting of files, folders, and models exceed the temporal capacities of archive staff. Proposals to alleviate such work by introducing AI pattern recognition tools for metadata genesis do not address this earlier process, as they assume ready-to-read files and digitized archivalia.

Digital archives face persistent challenges in findability and accessibility. These difficulties are shaped by internal structures and by the broader digital environment in which they operate. External factors make matters worse: search engines and recommendation algorithms often prioritize popular or commercial content, while pervasive advertising can obscure archival repositories. Typically hosted in isolated corners of the web, digital archives are organized according to unique, sometimes idiosyncratic, schemas that often lack standardization or metadata guidelines. As a result, they are rarely linked to or interoperable with other archives, which limits opportunities for discovery and cross-referencing. Even when archives adopt standards or ontologies such as Dublin Core or CIDOC CRM, there is significant variation in how data is modeled and interpreted. Archival records are shaped by technical frameworks, by institutional cultures, local practices, individual decisions, methodologies, AI tools, and evolving ideas about what should be preserved.

The same concept can often be represented in several valid ways, depending on historical precedent, disciplinary approach, or system limitations. Over time, these datasets tend to drift apart like languages developing in isolation. They fragment, merge, or are restructured, creating divergence even when they once shared a common standard. Initiatives such as Europeana and NFDI4Culture illustrate large-scale efforts to standardize access to cultural heritage data. To operate across diverse institutions, these platforms often depend on schemas that favor interoperability over contextual depth. This can lead to a flattening of complex, locally specific data. Formats, relationships, and metadata that do not align with the standard model are often omitted or oversimplified. The resulting loss stems from abstraction, standardization, and infrastructural constraint, which are not always intentional and may reflect what is feasible within available funding, maintenance capacity, or technical frameworks.

This brings us back to the question of archival loss in the digital age—not only what has vanished, but also what is systematically, temporarily, or unintentionally left out: an active act of losing rather than a passive one.

It might also lead to the question of what embodied knowledge is in the age of AI, and how these bodies and entities might learn to lose information, if they can.

Getting Lost: A Transitive Verb after All?

"The world, though, only grieves for what it knows, and has no inkling of what it lost with that tiny islet, even though given the spherical form of the Earth, this vanished dot could just as easily have been its navel, even if it was not the sturdy ropes of war and commerce that bound them one to the other, but the incomparably

finer-spun thread of a dream. For myth is the highest of all realities and—so it struck me—the library the true theatre of world events." [5]

If the archive is not a collection of *things,* but a co-living space of more or less animated matter, loss becomes a different category as well. Some of the loss is archivally desirable: dust cleaned from old slides, glue residue carefully scratched from paper, or duplicate copies of magazines or books that are donated or pulped. Experts track paperfish populations in archives through traps, meticulous counting and tracing their migration through books, rooms, shelves, and drawers. The negotiation between what needs to be shed to protect what is left versus the increasing awareness in archeology and related fields that any removal of "dirt" around "objects" has destroyed much of what one might have otherwise learned from matter stuck to the historical artifacts—in short, it disposed of yet to be decodable information. Loss here is both unattainable promise and dreaded threat. How can one read them together? And does loss always require a definition of that or who which loses, and that or whom that is lost? What predefined agency is necessary to be losing something, and what posterior awareness to call it that?

Neural networks are computational systems designed to mimic what was known about the brain's structure and function. Built from layers of interconnected neurons, they adjust the strength of connections based on input and output patterns. Through repeated exposure to data, neural networks reinforce certain pathways while diminishing others, forming internal models to classify, predict, and make decisions. As a result, learning in neural networks involves reinforcing important patterns, and forgetting or weakening those less rewarding based on the definition of the right answer. The human brain itself learns by strengthening useful synaptic connections and pruning those that are unused. In both systems, forgetting is not a flaw but an essential part of understanding, adaptation, and memory. Archives, in contrast, are not built to adapt in the systematic way of neural networks. However, over time, they do adapt. This adaptation tends to be porous, organic, and often unintended, unfolding within a dynamic ecology of loss and preservation. Archives are shaped by environmental conditions, institutional cultures, human and nonhuman agents, and evolving forms of access and use. Much like synaptic pruning in the brain or weight decay in neural networks, these processes of loss shape the archive's memory structure: what survives, what disappears, and how relationships between materials can be made. Archives are shaped as much by what they forget as by what they retain.

Likewise, the human brain is not a system built according to a fixed plan, but the evolutionary result of continual interaction with its environment, shaped by adaptation, exposure, trauma, and use. Brains evolve within specific bodies, cultures, and social entanglements. They host and are hosted by other entities, language, memory, emotion, habits, that emerge over time.

Much like archives and AI, the brain is not static but continuously restructured by what it encounters and what it forgets. Its intelligence arises not from completeness, but from its capacity to remain plastic, porous, and responsive to the environment. In this way, all three—brains, archives, and AI—are also constituted by loss and presence.

To date, AI, especially neural networks, are adopted in archives mainly to process existing materials, such as content, metadata, and document structures, focusing on what is already present. This includes automatic metadata extraction, document classification, entity recognition, similarity analysis, the construction of data networks, detection of recurring patterns, generation of transcriptions and translations, and the creation of dynamic visualizations of archival relationships. Clustering methods have supported provenance analysis, while centrality and network metrics have helped identify key figures or documents in historical datasets. These uses reflect an important and still growing role for AI in cultural heritage, automating labor-intensive processes and revealing previously unseen structures across large collections.

While these uses might reduce the human workload, critical studies on algorithm bias or metadata replicating colonial (or other) power structures have long pointed out the inherent problems of producing automated data from objects when the value systems and categories leading to their presence in the archive are always already imbued with discriminations, skewed assumptions, and biased observation[6]—be it through the skewed data pools impacting interpretation or creation of other data or metadata genesis from problematic object assemblies. Metadata is, still, the main interface of the archive; it is what renders the depths of the archive visible. Hence the chicken-and-egg problem of existing bias and generated data producing more bias are not only embedded in the archive systems, but feed from them and spawn them, both in data-generating AI solutions (be it for digitization or indexing), and sit on top of already existing bias (from data pools to stored objects). Loss, in this case, might be an alternative way to cut this hatching cycle.

While these critiques expose how biases are embedded in the data and AI systems, an equally urgent concern is what remains invisible to such systems.

Most AI approaches today focus on what is present, legible, and classifiable. There is a significant gap in algorithmic models and research that address what is missing, what has been excluded, degraded, or never captured in the first place. A shift is needed from merely optimizing AI for recognition, toward developing systems that can learn from incompleteness itself. This means designing AI not simply to operate on archives, but to grow with them, learning to sense and respond to what is absent, to identify patterns of silence, to trace discontinuities, and to understand the temporal, material, and curatorial conditions that shape access. This requires space for future engagement with both presence and absence, which is evolving rather than fixed categories. Practical strategies to support this approach could include metadata structures that model absence, enabling human and machine interpretation. Interface designs should help make missing or fragmented data visible, and training datasets should be constructed with ambiguity, uncertainty, and incompleteness in mind. With these tools, AI would not only support archival labor but participate in knowledge-making processes. It could help uncover, describe, and model absence as a meaningful part of archival systems. In doing so, loss would no longer be treated as a flaw, but as a generative structure that remains open for future interpretation.

Whether there once—soon—might be an AI that can not only simulate, but embody ambivalence, hesitation, doubt, and loss remains to be seen. For now we might need to start by understanding loss as the complex feature that it is, bugs and all.

1 Judith Schalansky, *An Inventory of Losses*, trans. Jackie Smith (London: MacLehose Press, 2022), p. 21.

2 María Puig de la Bellacasa, "Re-Animating Soils: Transforming Human–Soil Affections through Science, Culture and Community," *The Sociological Review* 67, no. 2 (March 1, 2019): 391–407.

3 Costanza Caraffa, "Manzoni in the Photothek: Photo Archives as Ecosystems, Prague 2017," in *Instant Presence: Representing Art in Photography: In Honor of Josef Sudek (1896–1976)*, ed. Hana Buddeus, Vojtěch Lahoda, Katarína Mašterová (Prague: Artefactum, 2017), pp. 121–37, here p. 134.

4 Puig de la Bellacasa, "Re-Animating Soils"; and María Puig de la Bellacasa, *Matters of Care: Speculative Ethics in More than Human Worlds* (Minneapolis: University of Minnesota Press, 2017).

5 Schalansky, *An Inventory of Losses*, p. 44.

6 Ruha Benjamin, *Race after Technology: Abolitionist Tools for the New Jim Code* (Cambridge, UK/ Medford, MA: Polity, 2020); Byron Ellsworth Hamann, *The Invention of the Colonial Americas: Data, Architecture, and the Archive of the Indies 1781–1844* (Los Angeles: Getty Research Institute, 2022); Mark C. Marino, *Critical Code Studies* (Cambridge, MA: MIT Press, 2020).

Giulia Bruno

Carrying Data

Automated tape library at Leibniz Supercomputing Centre (LRZ), used for long-term data archiving. Thousands of magnetic storage tapes are managed by robotic systems to ensure reliable, energy-efficient preservation of scientific data over decades. Tape storage remains a key component in large-scale research infrastructures due to its durability and low power consumption.
Leibniz Supercomputing Centre of the Bavarian Academy of Sciences and Humanities, Munich, Germany, 2025.

Carrying Data

In an era when data transmission becomes the basic premise of global communication—appearing both weightless and infrastructurally immense—the question of how we "carry" and "care for" data goes far beyond technological updates. It becomes a matter of collective memory, justice, and who holds the right to tell history. What does it mean to safeguard data not only on servers or hard drives, but also in the architecture of the world, in geology and stone, in the rhythms of plants, and in the guardianship of seeds and biodiversity patents?

Can we think of nature itself as an archive, where the act of carrying includes erosion, sedimentation, oblivion, and regeneration?

Archiving is never a neutral act. The spaces where data are stored—data centers, glacial archives, and underground servers—are often invisible to the human eye and designed according to both human and nonhuman logics, optimized for technology.

They are architectures of opacity. The "cloud" is inseparably tied to the artificial transformation of territories, and digital storage carries a material violence: rare earths extracted from formerly colonized lands, energy generated from extractive economies and relations of dominance.

Caring for data means acknowledging its physicality. Data has weight, temperature, and a territorial footprint. It lives in circuits, but also in languages, in protocols, metadata, and formats we must constantly update to avoid obsolescence. Without reading tools, without contemporary Rosetta Stones, vast bodies of knowledge risk becoming inaccessible. Technology is not merely a tool; it is a language. And just like lost alphabets or erased oral traditions, losing technological legibility means losing entire worldviews.

Who holds the keys to the archives of the future? What epistemologies are privileged in the devices that store memory? Like colonial libraries, digital infrastructures can reproduce dynamics of domination and exclusion. Access to data is often conditioned economically, linguistically, and geopolitically, just like access to the energy, mineral, and geological resources on which these technologies rely. It is in this relationship that data is revealed not just as symbol, but as political matter.

The climate crisis, the decisions debated at the United Nations Climate Change Conferences (COPs), and the mechanisms of global environmental governance all show how access to resources is never distributed equally. Even data, including climate data, is extracted, measured, owned. Observing the planet from space, through satellites and environmental surveillance tools, exposes a form of immaterial extraction with deeply material, politically determined consequences.

Contemporary physics from the underground labs of Gran Sasso to emerging quantum research invites us to consider a new scale of observation, where the macro and micro mirror each other. Scientific infrastructure becomes an archive itself: of particles, of radiation, of probabilistic models that attempt to read the invisible. Here, data becomes language, and language becomes filter: what we can observe increasingly depends on the tools we use to observe. And these tools are themselves loaded with political, economic, and aesthetic decisions.

In this context, nature is not just metaphor: it is co-author. From ice core samples to tree rings, from ocean sediments to the salt crusts of ancient lakes, nonhuman archives bear witness to time with a precision that technology can only strive to emulate. But to read them, we need instruments and also trust in other forms of knowledge: scientific, poetic, Indigenous.

What does it mean to consider natural matter part of a shared archive? And what does it mean to carry (and care for) data not only through servers, but also through interspecies guardianship?

Ultimately, data is matter. And like all matter, it is contested. To carry it is to take part in the never-ending story of who gets to remember, speak, and narrate time. Multiplying perspectives, preserving obsolete formats, listening to both silence and noise: these are acts of care. And they are acts of resistance. The archive, human or nonhuman, is not only a site of preservation but a battleground for the right to tell time and history from multiple voices, positions, and worlds.

Robotic arm operating within the IBM TS4500 tape library at LRZ. Designed for high-density, automated data archiving, this system uses LTO-9 cartridges (up to 45 TB compressed) to ensure scalable, long-term storage of scientific datasets. The robotic handler retrieves and inserts tapes without human intervention. Leibniz Supercomputing Centre of the Bavarian Academy of Sciences and Humanities, Munich, Germany, 2025.

IBM TS4500 tape library at LRZ, based on LTO-9 technology. As the next-generation system for long-term data archiving, this robotic library supports storage capacities of up to 45 TB per cartridge (18 TB uncompressed). Data is currently being migrated from the previous tape system to ensure continuity and expanded capacity for large-scale scientific datasets. Leibniz Supercomputing Centre of the Bavarian Academy of Sciences and Humanities, Munich, Germany, 2025.

Photograph of the ACS4400 tape library used at LRZ from 1995 to 1999. This automated storage system—
"Speicherbibliothek ACS4400"—was part of LRZ's early efforts to manage large-scale scientific data archiving with
robotic media handling and high-density tape storage.
Leibniz Supercomputing Centre of the Bavarian Academy of Sciences and Humanities, Munich, Germany, 2025.

Control Data Cyber 175 mainframe in use at LRZ in the early 1980s. This photo shows the operation of one of LRZ's
large-scale systems of the time, with magnetic tape storage units and removable disk drives in the background.
The Cyber 175 marked a key stage in the center's transition to interactive and high-throughput scientific computing.
Leibniz Supercomputing Centre of the Bavarian Academy of Sciences and Humanities, Munich, Germany, 2025.

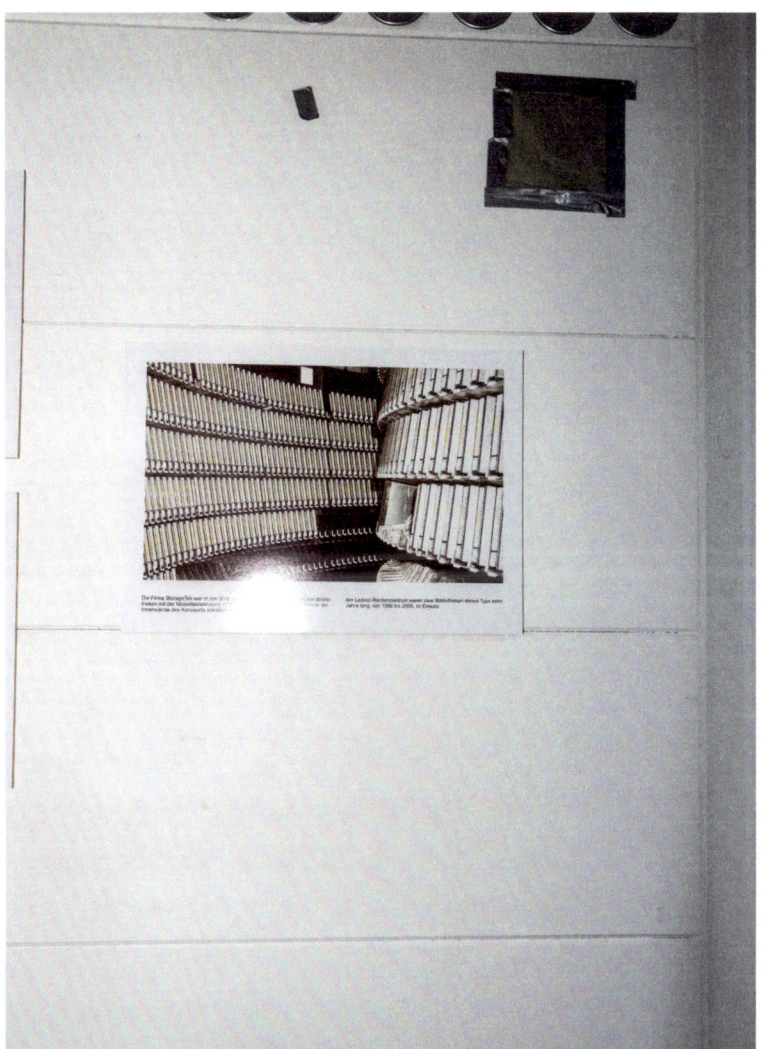

Historic image of magnetic tape storage units used in LRZ's early data archiving systems. Magnetic tape libraries were central to long-term storage strategies throughout the twentieth century, offering scalable, cost-effective solutions for safeguarding large volumes of scientific data. Leibniz Supercomputing Centre of the Bavarian Academy of Sciences and Humanities, Munich, Germany, 2025.

Magnetic tape drives in operation at LRZ, likely late 1970s to early 1980s. These large-format reel-to-reel tape systems were central to data storage workflows, allowing sequential access to scientific data. Operators managed tape loading, retrieval, and system maintenance manually.
Leibniz Supercomputing Centre of the Bavarian Academy of Sciences and Humanities, Munich, Germany, 2025.

Decommissioned LTO data cartridges from LRZ's previous tape library system. The cabinet contains LTO-7 tapes (top shelves) and older LTO-6/5 formats (bottom, black cases). As the system phases out, these cartridges are securely collected and sent for certified destruction to comply with data protection regulations. Leibniz Supercomputing Centre of the Bavarian Academy of Sciences and Humanities, Munich, Germany, 2025.

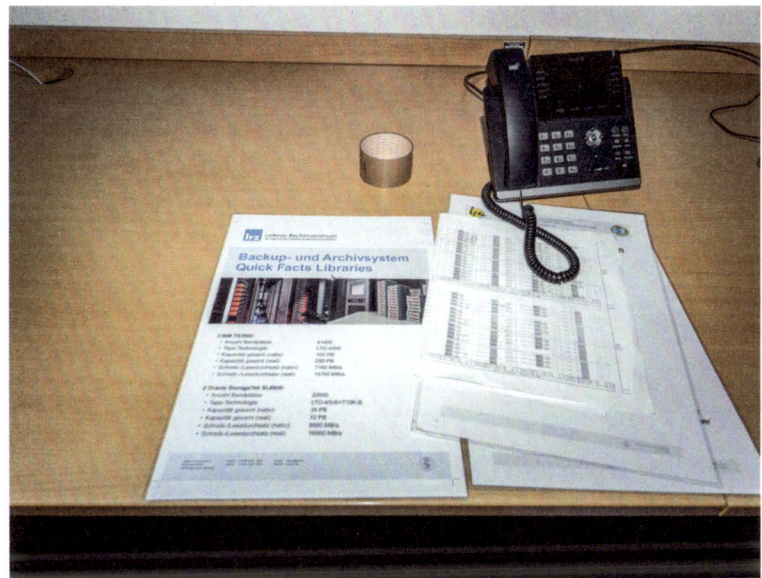

Workstation with documentation on LRZ's backup and archive systems. The printed "Quick Facts" sheet outlines technical details of the data tape libraries, while the phone and spreadsheets support daily operations and monitoring tasks related to secure data storage and archiving.
Leibniz Supercomputing Centre of the Bavarian Academy of Sciences and Humanities, Munich, Germany, 2025.

Basic workstation setup within the LRZ data center. Equipped with standard desktop hardware and a telephone for internal communication, this desk supports operational monitoring, troubleshooting, and coordination tasks within the high-performance computing environment.
Leibniz Supercomputing Centre of the Bavarian Academy of Sciences and Humanities, Munich, Germany, 2025.

Facility management clipboard mounted in the infrastructure ring of the LRZ data center. Located near an external air vent in the middle "onion" ring, this area houses essential building services. The clipboard is typically used by facility staff to document maintenance tasks or chemical safety information relevant to operations.
Leibniz Supercomputing Centre of the Bavarian Academy of Sciences and Humanities, Munich, Germany, 2025.

Temporal

Table with deconstructed LTO cartridges and a tape drive unit at LRZ. This image shows components from various generations of linear tape-open (LTO) technology. The older library system at LRZ used LTO-6 and LTO-7 (2.5–15 TB per cartridge), while the current IBM TS4500 system employs LTO-9 cartridges with capacities of up to 45 TB (compressed).
Leibniz Supercomputing Centre of the Bavarian Academy of Sciences and Humanities, Munich, Germany, 2025.

Backside of the tape library system in LRZ's dedicated backup and archive room. This secured area is exclusively used for long-term and redundancy-focused storage systems. The open cabinet reveals the internal structure of the tape library's robotics and power distribution units, supporting automated data archiving and retrieval. Leibniz Supercomputing Centre of the Bavarian Academy of Sciences and Humanities, Munich, Germany, 2025.

Temporal

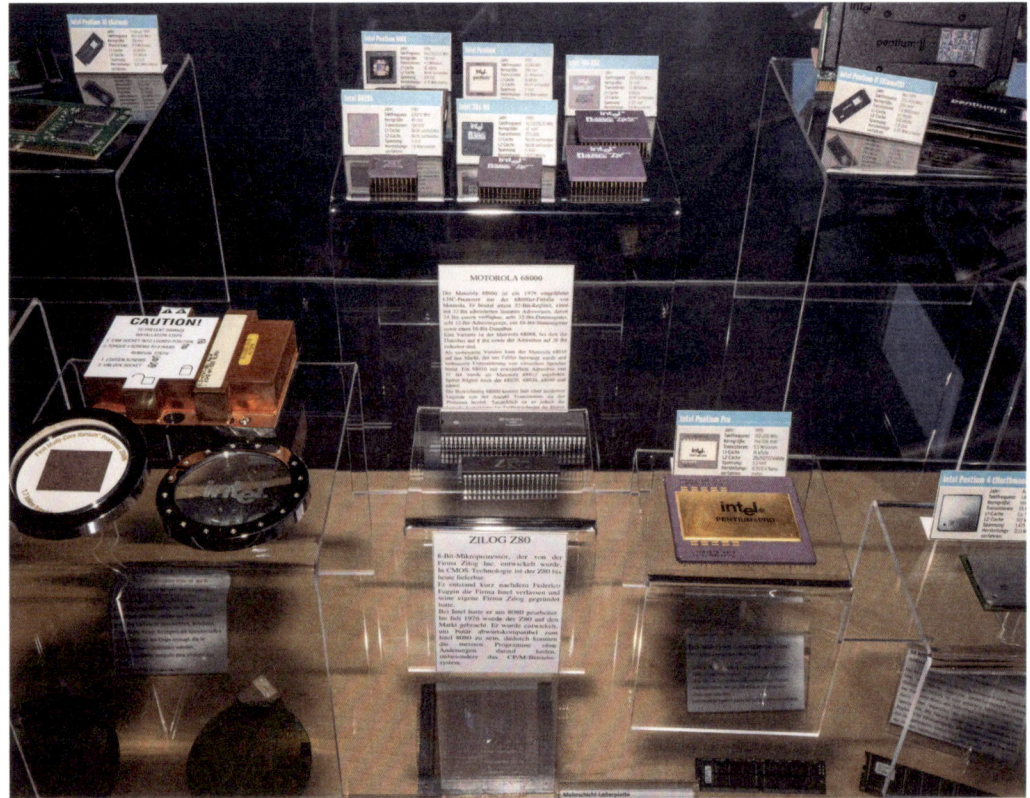

Processor and chip technology showcase at LRZ. This curated display traces the evolution of microprocessor and memory hardware—from early silicon wafers to high-performance computing components—highlighting key milestones in the development of scientific computing infrastructure.
Leibniz Supercomputing Centre of the Bavarian Academy of Sciences and Humanities, Munich, Germany, 2025.

Data Science Archive at LRZ, part of the center's long-term storage infrastructure. Designed to securely preserve vast amounts of scientific data, this archive supports research reproducibility, long-term access, and interdisciplinary collaboration across domains such as climate science, astrophysics, and biomedicine.
Leibniz Supercomputing Centre of the Bavarian Academy of Sciences and Humanities, Munich, Germany, 2025.

Exterior view of the LRZ campus in Garching, showing the Twin Cube (center) and surrounding buildings. The data center and office wing on the left were inaugurated in 2006, while the institute building on the right was completed in 2012. Together they form the architectural core of LRZ's operations in high-performance computing, research, and data infrastructure.
Leibniz Supercomputing Centre of the Bavarian Academy of Sciences and Humanities, Munich, Germany, 2025.

Temporal 171

Multiscreen control room for live signal management and global video conferencing during high-level climate negotiations. The setup supported real-time coordination and broadcast of sessions at the 2015 United Nations Climate Change Conference (COP21).
Paris, France, 2015.

Media center with public computer terminals and workstations supporting journalists and delegates. This digital infrastructure enabled real-time communication, content production, and data access during the 2015 United Nations Climate Change Conference (COP21). Paris, France, 2015.

Plenary hall prepared for high-level negotiations and public sessions. The modular structure hosted heads of state, diplomats, and observers during the 2015 United Nations Climate Change Conference (COP21). Paris, France, 2015.

Workstations and printing infrastructure within the media and delegation support area. This logistical environment enabled document production, communication, and coordination during the 2015 United Nations Climate Change Conference (COP21). Paris, France, 2015.

Temporal

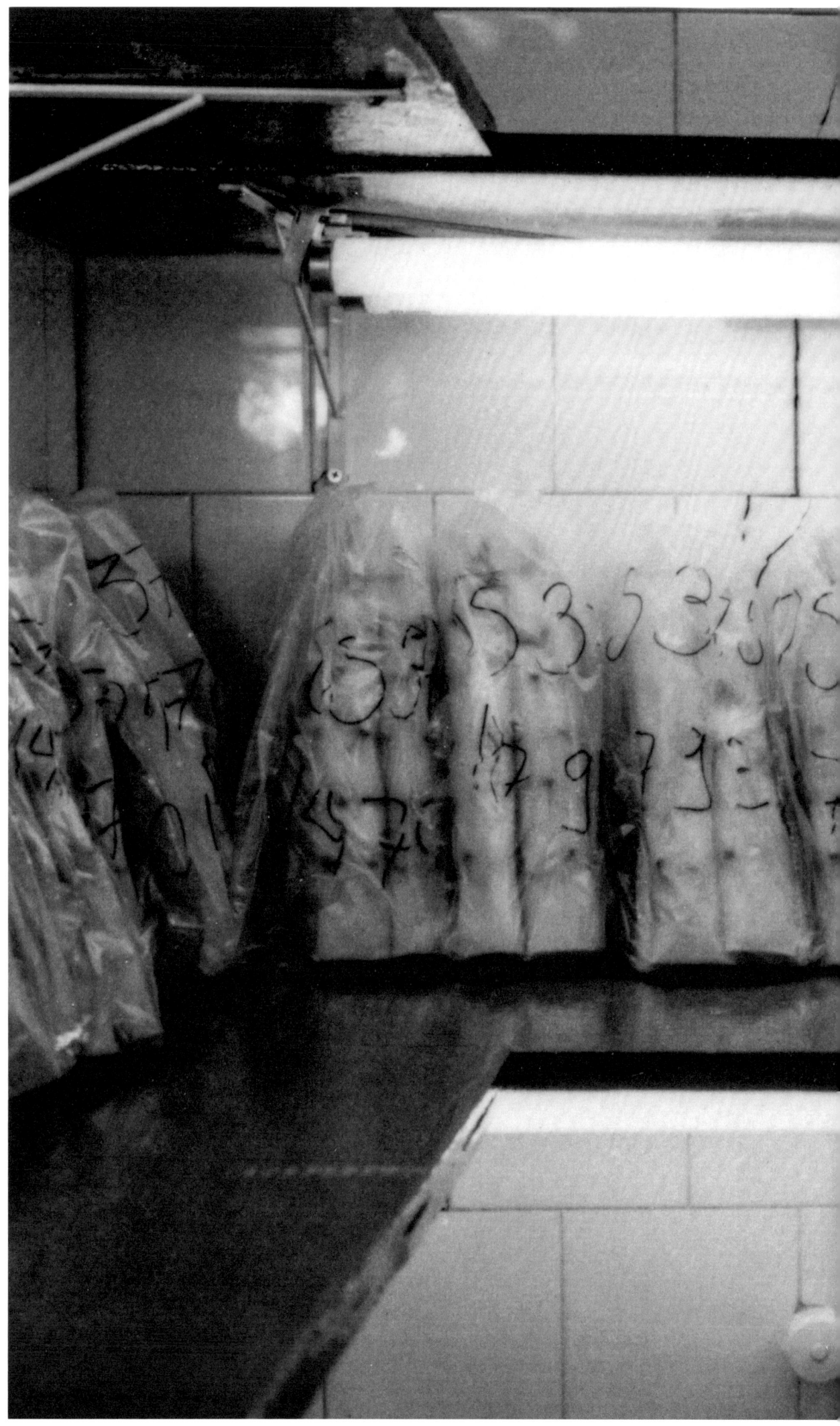

Labeled sample bags stored in a controlled environment for germination testing or genetic analysis of soybean strains. Such archival practices are essential in monitoring seed quality and supporting industrial-scale soy production.
Soya Lab, Sinop, Brazil, 2015.

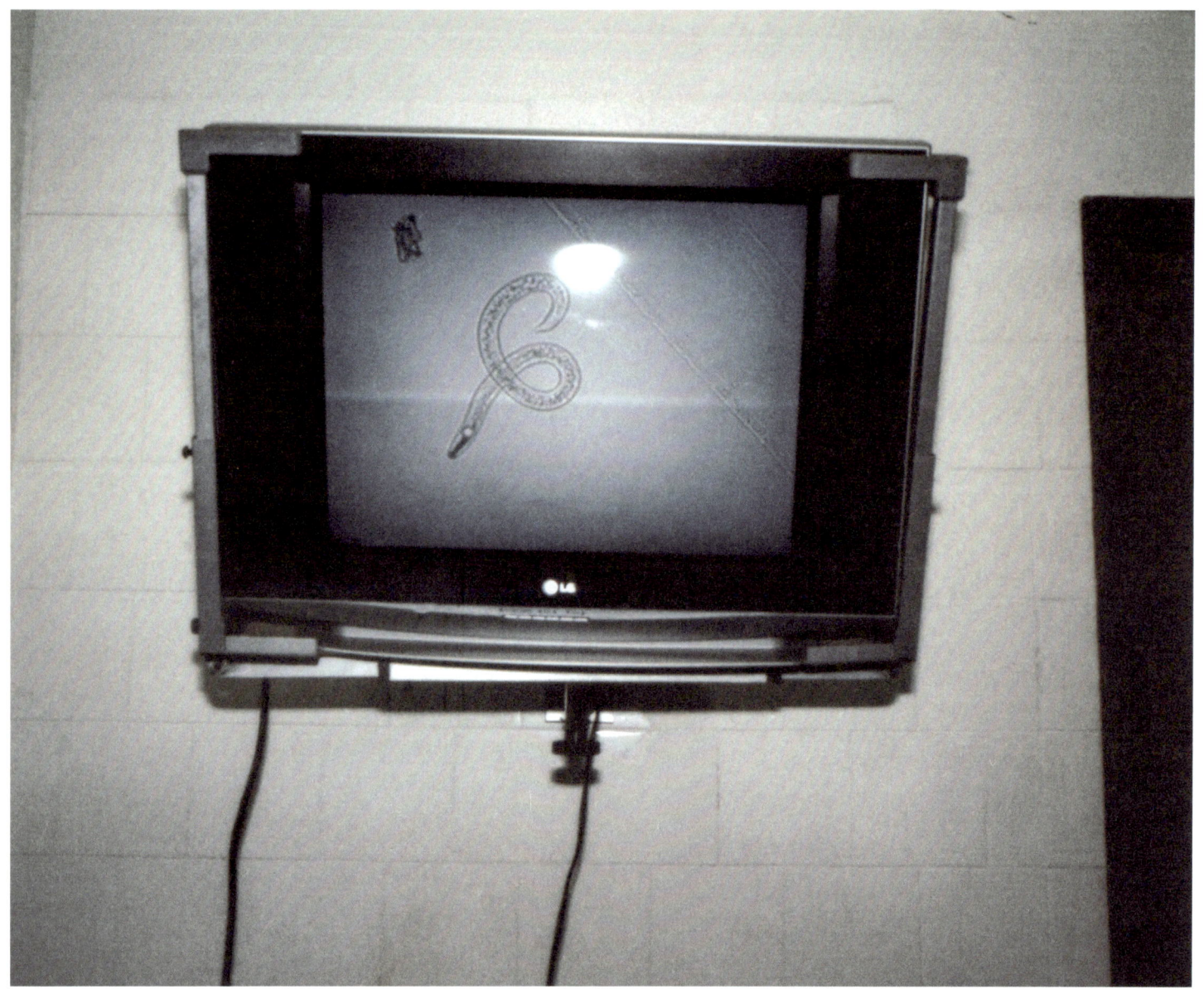

Laboratory monitor displaying imaging data related to soybean analysis and quality control. These visual diagnostics are part of agribusiness laboratory practices within soy production networks.
Soya Lab, Sinop, Brazil, 2015.

Flooded forest landscape typical of the Amazon biome, where palm species dominate the riparian vegetation. These ecosystems are critical for carbon storage, biodiversity, and water regulation, yet increasingly threatened by agricultural expansion.
Sinop region, Mato Grosso, Brazil, 2015.

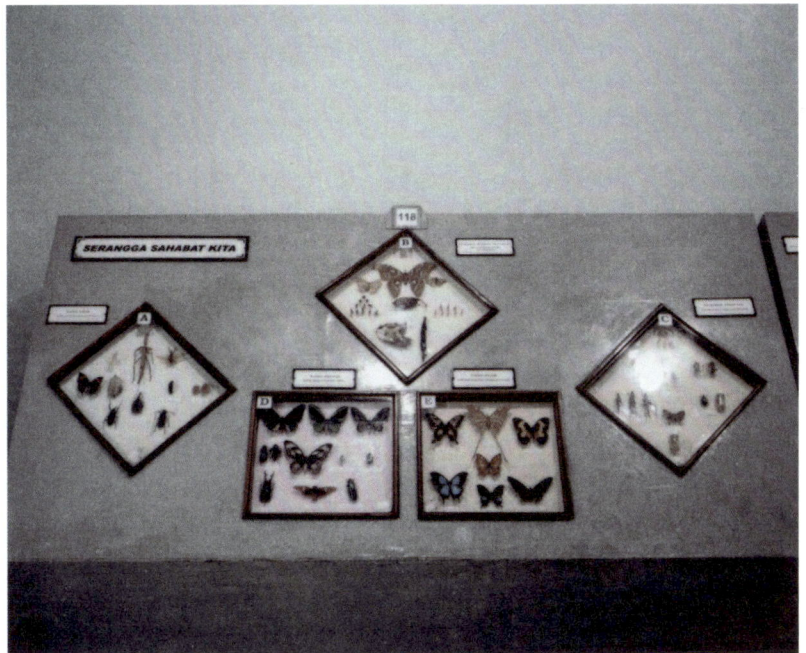

Entomological specimens from the Indonesian archipelago, displayed as part of the natural history collection. The Bogor Botanical Garden, established in 1817 under Dutch colonial administration, served as a central hub for biological classification and colonial scientific exploration.
Bogor Botanical Garden, Indonesia, 2016.

Botanical specimens preserved in liquid solution as part of the Cibodas Botanical Garden's archival collection. Originally established in 1852 during Dutch colonial rule, the garden served as a site for plant acclimatization and taxonomic research within imperial scientific networks.
Cibodas Botanical Garden, Cipanas–Cianjur (Jakarta), Indonesia, 2016.

Temporal

Control panel with protected keyboard and voltage warning signage, part of the technical infrastructure supporting scientific experiments.
Gran Sasso National Laboratory (LNGS), L'Aquila, Italy, 2021.

Thermal and acoustic insulation layers forming part of the Borexino experiment's shielding system. These materials help reduce environmental noise, enabling the detection of low-energy solar neutrinos.
Gran Sasso National Laboratory (LNGS), L'Aquila, Italy, 2021.

Technical infrastructure of the XENON experiment, supporting the operation of a liquid xenon detector designed to search for dark matter particles.
Gran Sasso National Laboratory (LNGS), L'Aquila, Italy, 2021.

Structural enclosure of the XENON experiment, which searches for dark matter using a liquid xenon time projection chamber shielded deep underground.
Gran Sasso National Laboratory (LNGS), L'Aquila, Italy, 2021.

Temporal

Borexino experiment, detail of the shielding and internal components. Designed to detect low-energy solar neutrinos, Borexino is an ultra-sensitive liquid scintillator detector operating deep underground to minimize background radiation.
Gran Sasso National Laboratory (LNGS), L'Aquila, Italy, 2021.

Collection of scientific publications archive.
Gran Sasso National Laboratory (LNGS), L'Aquila, Italy, 2021.

Galaxy poster with a relativistic jet.
Gran Sasso National Laboratory (LNGS) L'Aquila, Italy, 2021.

"History of the Universe" poster.
Gran Sasso National Laboratory (LNGS), L'Aquila, Italy, 2021.

Temporal

Ceramic casings for sodium-nickel battery cells.
FZSONICK S.A., Stabio, Switzerland, 2023.

Ceramic components during the firing process in a high-temperature kiln.
FZSONICK S.A., Stabio, Switzerland, 2023.

Interior of a high-temperature tunnel kiln used for ceramic component processing.
FZSONICK S.A., Stabio, Switzerland, 2023.

Sorting station for battery components by size (small, standard, large).
FZSONICK S.A., Stabio, Switzerland, 2023.

Temporal

Satellite laser ranging (SLR) telescope housed in a protective dome, used for high-precision distance measurements to geodetic satellites. In the background, a large parabolic antenna supports VLBI and deep space tracking. The facility contributes to international networks including ILRS, IVS, and IGS.
ASI – Italian Space Agency, Space Geodesy Center, Matera, Italy, 2019.

Ground-based antennas forming part of the Space Geodesy Center's tracking infrastructure. These systems support satellite laser ranging (SLR), GNSS signal acquisition, and deep space communication, contributing to global networks such as the International Laser Ranging Service (ILRS), the International GNSS Service (IGS), and the International VLBI Service (IVS).
ASI – Italian Space Agency, Space Geodesy Center, Matera, Italy, 2019.

Office workstation with printed imagery of LAGEOS (laser geodynamics satellites), used in satellite laser ranging for high-precision Earth measurements. These missions contribute to geodesy, tectonic plate monitoring, and global reference frame maintenance.
ASI – Italian Space Agency, Space Geodesy Center, Matera, Italy, 2019.

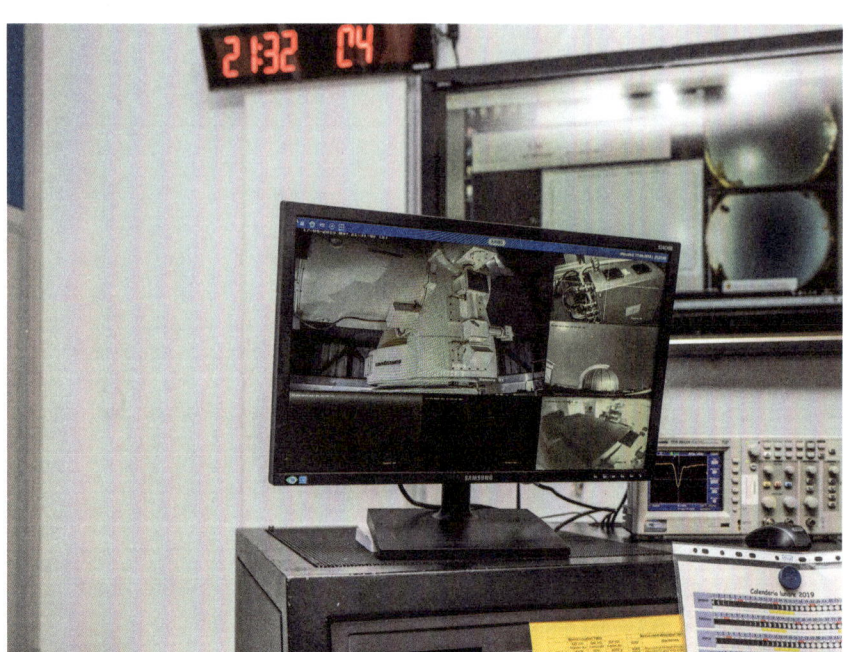

Surveillance monitor displaying multiple camera views of satellite tracking and instrumentation systems. These visual feeds support the remote monitoring of antenna operations and equipment status.
ASI – Italian Space Agency, Space Geodesy Center, Matera, Italy, 2019.

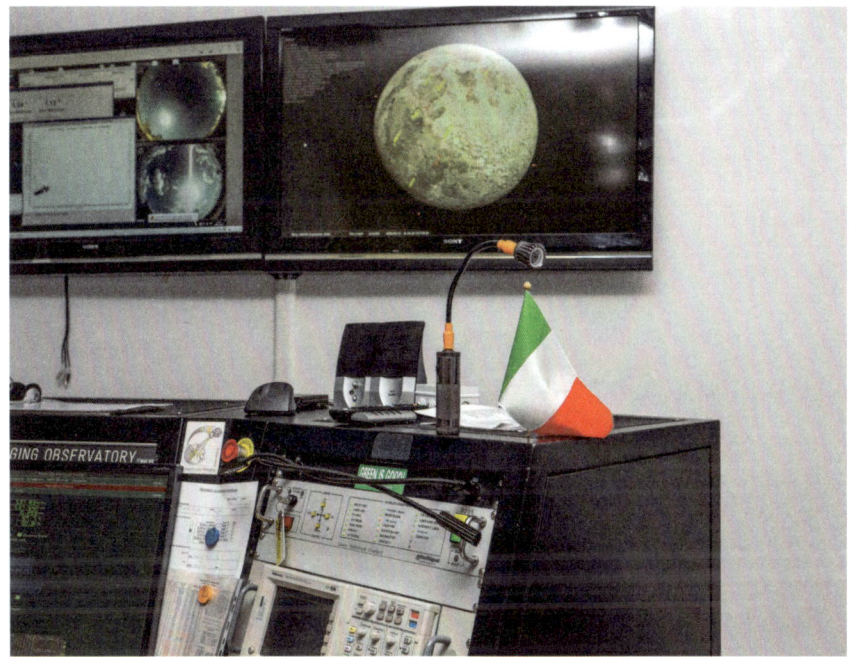

Instrumentation panel and monitoring displays used for satellite data acquisition and lunar observation. The Space Geodesy Center plays a key role in global geodetic networks and deep space tracking operations.
ASI – Italian Space Agency, Space Geodesy Center, Matera, Italy, 2019.

Control workstation displaying real-time satellite tracking and lunar observation interfaces. The Space Geodesy Center supports high-precision measurements for Earth observation, orbital dynamics, and interplanetary missions.
ASI – Italian Space Agency, Space Geodesy Center, Matera, Italy, 2019.

Calibration and instrumentation lab containing legacy and active measurement equipment used for geodetic and space tracking systems. These tools support the maintenance and precision alignment of sensors contributing to international geodesy networks.
ASI – Italian Space Agency, Space Geodesy Center, Matera, Italy, 2019.

Data storage units labeled for formatting and verification, used for archiving satellite observation and telemetry data. These physical media form part of the long-term data management strategy supporting geodetic and space missions.
ASI – Italian Space Agency, Space Geodesy Center, Matera, Italy, 2019.

Epilogue

James Bridle in Conversation with Cara Hähl-Pfeifer and Damjan Kokalevski

Regaining Agency in a Technological Swamp

In conversation with the editors, writer, artist, and technologist James Bridle expands on ways to regain agency vis-à-vis persistent forms of technological dispossession and abstraction. Bridle discusses agency as a starting point in accessing other forms of intelligence, while outlining a form of practice to understand systems of planning, practice, and ecologies surrounding us. From cybernetics to slime molds and computation, to James's self-built practice, a bold idea for ecological thinking emerges.

Cara Hähl-Pfeifer and Damjan Kokalevski: In *Ways of Being* (2022), you introduce various more-than-human intelligences, such as the case of *Physalum polycephalum,* a type of slime mold, which very efficiently recreated the rail system of Tokyo. This task is a notoriously difficult mathematical problem, as you describe it, because the number of possible solutions increases as more regions and cities are added to a system. And yet the mold managed to solve the network in linear time. What do you think are the benefits and value of these other forms of spatial intelligence over computer models?

James Bridle: The slime molds are a great example.[1] They show their capability of maneuvering around a set of complex constraints. But what's really particular about the slime molds is trying to understand the efficiency by which they came to those realizations, which reveals an entirely different way of thinking about problems. We humans might reach the same answer, but how quickly we get there and in what way is an interesting computational question. The generalized problem is called "the traveling salesman problem," a question of how you get between various points in the most efficient manner. The traveling salesman says: If you've got six cities, what's the shortest route to visit all of them only once, and not retrace your steps? That's an incredibly hard problem, because mathematically, there's this exponential number of possibilities: the only way to do it is to calculate every single distance. That means as you add more cities, the problem gets bigger, and the time to reach a solution increases exponentially. But when they set this problem for slime molds, the molds solved it in linear time, and we simply don't know how they do it. This just completely baffles all our ways of thinking, because both humans and digital computers find this problem essentially insolvably hard.

And that's the second interesting thing: how the slime molds think about this is essentially inscrutable to us. We don't know how they do it. We can biologically go into some of the processes that are involved. Yet, the question arises: How do we engage with processes where we don't fully understand the mechanisms involved, but we have to work with systems that are capable of solving them? That brings us back to computation, where we're frequently working with incredibly complex technological systems that very few of us fully understand, and really no one has a complete overview of. So, there is always this kind of level of trust of engaging with complex systems, and we have to decide, to what extent it is morally, ethically, and intellectually necessary to have a complete overarching understanding of a thing, or whether there are different forms of cooperation possible. Ultimately, what kind of thinking are we doing here, when so much of particularly Western intellectual scientific method thinking has always been about knowledge as a form of domination, where we understand things in order to take their power and use it for ourselves. One of the things that we have learned from nonhumans is that these processes are so inaccessible that we can either ignore them, we can carry on the kind of deeply extractivist process, which is always reductive and is never going to yield the full benefits, or we can learn new forms of cooperation.

What you just described, the idea of not knowing what this intelligence is, or not knowing how to access it, would be every spatial planner's nightmare because the idea in planning is that you create infrastructures and cities in order to run them as efficiently and controlled as possible. This all comes, as you pointed out, from our Western ideals of science being quantifiable and applicable. In contrast to these, can you outline alternative ways of accessing these other types of intelligence that you described?

If we offload these tasks, this planning, in daily operation, to systems that we don't have full control over, then where does accountability rest? We have to remember, when there are health and life and various other factors at stake, then it does matter how much understanding we have. At the same time, some sense of possibility is lost. We already employ systems akin to the slime mold in limited ways that are within our control. I'm thinking of the clams in the water filtration systems in Poland,[2] which are wired up to sensors. The clams have much more sensitivity to pollutants within the water than anything that humans have developed. As the water passes through, if the clams close up, the water is cut off immediately. That's an example of where these systems are integrated. And I think the lesson in that is that the integration, whatever form it might take, comes from awareness and paying attention to the actual life of other beings, even though much of the intelligence at work may be unknowable to us.

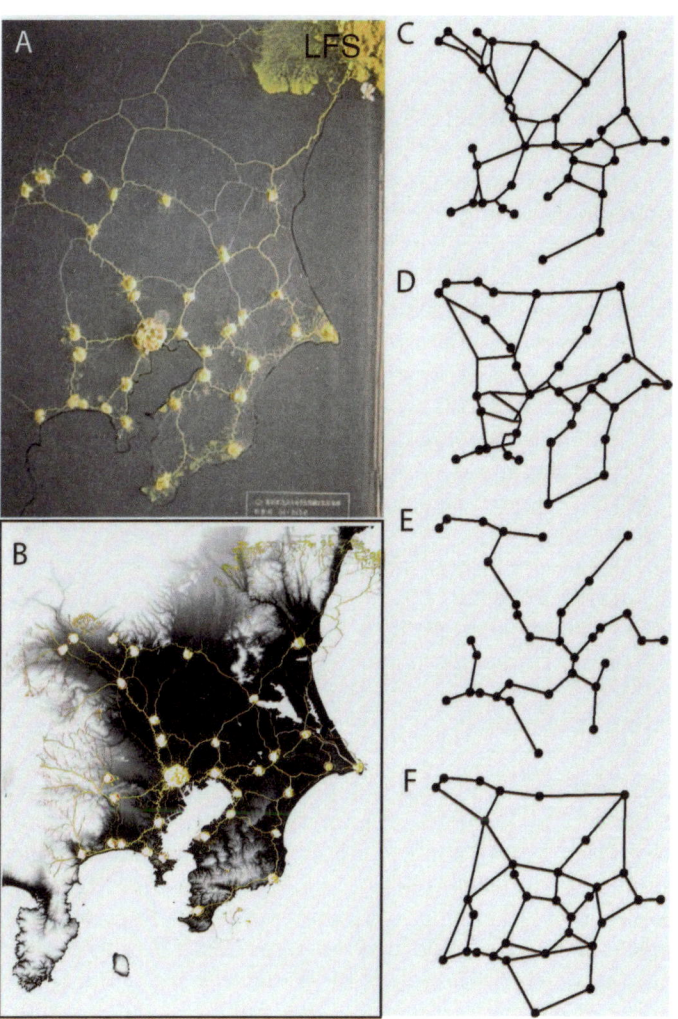

Comparison of the slime mold networks with the Tokyo rail network.

Epilogue

> One of your concerns in *Ways of Being* is cybernetics and how technology can adapt to its environment. It can offer us a relational, agential way of thinking. What might be the relevance of cybernetics for spatial planning and the environment?

I'm mostly interested in some of the weirder British cyberneticians working in the 1950s and 1960s who had very strange ideas about incorporating nonhuman intelligence into ways of thinking. In the book, I write about Stafford Beer, for example, who, in attempting to design intelligent systems, realized that you couldn't build an intelligent system from scratch. An intelligent system had to be living, aware, and connected to the world. The reflection from that area of cybernetics is an understanding that intelligence is something that's an active process, and not reducible to a black box at all. For any system to act with the world, it needs to be firstly connected to that world, and secondly, responsive to it—changing in response to its environment, to events that occur. It is itself alive in some function, and therefore, intelligence is relational. It exists between things rather than just inside boxes. So, with regard to artificial intelligence, you can see this evolution happening both at the macro scale and the micro scale. For example, at a micro scale and within cities, you see this happening with autonomous vehicles. When they started with the original ideas for autonomous vehicles, you would have a computer filled with the rules of road driving and a map of the city. That completely failed, because cities and geography are way more complex than that. Instead, today these systems are trained, and essentially any autonomous vehicle that's out on the road has already driven both physically and virtually, hundreds and thousands of miles in order to learn what it is like to drive in the real world. So, there's this deep understanding that emerges from those computational practices, but which should already be obvious from environmental practices. Intelligence is something active and out in the world; it is an ongoing process of learning and acting, and that, for me, is at the core of the cybernetic thinking that I'm interested in.

> When we design a system or infrastructure, we think of how much is fixed and how much is responsive in daily practice. In architectural conversations nowadays, we discuss how cities and homes need to become smarter. One way to see this through is by adding more sensing capabilities and advances so that we can create autonomous or semi-autonomous systems that learn from these inputs. In a way, it's like we are living inside one big computer. And so maybe the smart city, the smart home, is the epitome of this endless stream of optimization. Can you talk about the dependency on technology and the problem of automation bias in relation to the smart city and the home?

Automation bias is a really solid and terrifying finding from research into human-computer interaction. In my first book *New Dark Age* (2018), I write about this example of where they put very experienced pilots into a simulator and gave them an automated system that would give advice at different times, which they could choose to follow or not. It was discovered that when the system gave these experienced pilots advice contrary to their training, which would be dangerous in the situation they found themselves in, the vast majority of the time, they followed the automated device rather than the training. And this seems to be a real sort of cognitive effect, almost a hack, I would call it. Our brains are optimized to take the most efficient approach to anything, to expend the least energy, to act on the information it is given. When you have a system that presents itself authoritatively in this way at a critical moment, the brain defers to that information, and that is a huge problem for all of our interactions with technology. We subconsciously, sometimes consciously, trust these systems. Most of the time, we can, but because there are times when we shouldn't, it slips right under the radar, and we follow the instructions of these systems all the time. The same applies to following navigation on Google Maps. These technological systems may have been designed by humans, but those decisions are made or channeled within the systems, and not necessarily with the intentions of the original programmers.

> In architecture and building industries, over the past few decades, the design process in CAD software has been automated and streamlined by what is known as "building information modeling," or BIM: normed and standardized 3D objects, such as windows or doors, that you can select from a drop-down menu and use to design your building without asking how they came to be. These objects also include all the metadata and details you need for the construction phase. In this way, you literally create a digital twin of your building within an integrated workflow.

So much of my thinking about this started from a chance encounter with an architect a decade or two ago. I'd become obsessed with a certain style of architecture that I saw appearing around London, where I lived at the time. A certain kind of new-build mass-housing architecture that all really looked the same. When I asked a couple of architects and planners about this, the explanation that came back was that these buildings all look the same because everyone uses the same software to develop them. And you could see that there are default options inside any of these software programs. And most of the time, people stick to the defaults. This is not in itself a terrible thing. Those defaults are probably there for a reason, and quantified around a whole bunch of useful metrics and so on. But ultimately, you start asking the question: Where is agency within the system? Who designed this building? Was it the architect or the engineer who designed it? Was it the person who designed the software? Was it the software system itself? I don't think we're always aware of the scale at which this operates across pretty much all of society and culture. And critically, that most of our systems are designed to obscure this effect. Certain kinds of computer interactions should be more frictional. When we're talking about urban scale systems, people should know who or what is accountable for the things that they interact with, where to address complaints, and critically, how to make their own changes. How they, as users, inhabitants, and citizens of spaces, can engage with that. Therefore, transparency and essentially friction within these systems are far more necessary than most of our design methodologies.

You just mentioned something that's quite interesting. In the postwar period, a lot of buildings, mainly office buildings, were built in such a way that you could no longer manually open the windows because of fully automated HVAC systems, like air conditioning, for example. And I don't know if you've ever experienced this, but it can feel quite suffocating not to be able to open the window. There is even a name for this psychological-turned-physical discomfort, ironically called the "sick building syndrome," depriving people of any agency. Do you believe we can think differently about smartness and agency?

I get a very strong sense that it is also a psychological response to being within spaces where one does not have agency. And agency for me is one of these things that I keep coming back to as a kind of foundational necessity and also an explanation for so many of the situations that we find ourselves in, in the present moment. For example, it explains a huge amount about contemporary politics around the climate. Our agency with respect to the climate crisis is affected by the utter lack of leadership by government and the antagonistic behaviors of large corporations, our own deep-rooted need for energy and high living standards, and so forth. But there is also a lack of action on our own part, because this problem feels so vast and incomprehensible. It just feels totally hopeless. We become disengaged. How do you flip that over and start to restore agency?

I've found within my own practice that my deep feelings of depression and trauma regarding the future in general and the climate in particular are closely related to this lack of agency. And so, I take action to address this by learning about things. I'm currently building a house, and in order to do that I had to go through a huge process of planning applications in which I had to specify in advance down to every single screw, nut, bolt, and beam, exactly how this house would look at the end. I'm working on the framework of the architect Walter Segal, a German-British architect whom I think many architects are familiar with, even if they're not able to follow all of his dictates.[3] But Segal fundamentally believed that, first of all, anyone could build their own house, something which I am currently stress testing. But also that this produced a better kind of architecture, because you weren't locked into a certain narrow vision of an outcome that came from planning law and certain types of architectural process. The way in which Segal designed his basic framework, those buildings are highly alterable by generations of inhabitants. They're very easy to change, which is not the case with most modern buildings. But for architectural reasons, for reasons of material choices, also for reasons of planning law, all these things affect the extent to which these buildings are malleable. One of the reasons I wanted to build my own house was getting really fed up of living for decades in rented properties where I was constrained from doing what I wanted to do to the building itself, because I didn't legally own it, but also because I had no idea what was behind the walls. One of my great architectural inspirations is Eileen Gray's E-1027 Villa. If anyone's been there, they'll remember she has this extraordinary surface-mounted wiring system throughout the entire house. This idea of surface mounting visible systems has appeared in architecture at various times—obviously the Pompidou Centre, the Lloyd's building—and with this externalization of functions is for me probably more of a computational-technological example than it is an architectural one. I want to live in a house where I can see all the systems and I know where all the wires are.

A movement of low-tech, regional, and simple building is reemerging in today's generation of architecture students and young practices. In recent years, the Technical University of Munich has offered several workshops where experts shared some of their knowledge and skills with us students on how to work with regionally sourced and locally produced materials. It is workshops like these that largely contribute to empowering architecture students, something I have experienced first-hand. It should not be this complicated to build a building, but it is the contemporary canon.

I can't stress enough, because I sometimes feel silly talking about this, that it's a surprise to me as well. It is so opposite to the culture that we've been raised in. One of the other things I'm doing in the house that I'm building is a large part of the insulation. I'm using *Posidonia oceanica,* which is the sea grass that grows in the ocean. It's what washes up on the beaches all around the Mediterranean. And for as long as humans have been around here, that material has been used for bedding and for insulation. If you walk around the island that I live on, you see old buildings, ruined buildings, where the stone walls are stuffed with this leaf matter. It's an incredibly good material. Beyond being just sustainable and free to collect on the beaches, it's also basically fireproof, pest-proof, and very good thermal insulation. But it also reconnects with the environment in a way that I think is both beautiful and necessary. Because through my engagement with *Posidonia*, I've been drawn into the sea, and spend time swimming among the living off the shore. And I've become involved in seagrass conservation. It makes a link between the materials we use, the source of those materials, the ways in which they're extracted, in a way that makes us pay attention and ultimately, hopefully, care in some way, even selfishly, because if I want to keep building with seagrass, I have to make sure the seagrass lives. I think it's really important to go beyond just thinking of these materials as abstract goods and understanding what their use really does for us to create more regenerative, spiritually nourishing, and enriching living conditions for all.

It's so interesting that we started by asking, how can we access this knowledge? You just described a way of basically discovering the environment around you by using not just your mind, but also your body and its interaction with its surroundings. This is obvious, but may sound easier than it actually is to implicate.

Walter Segal method under construction in 1988.

Publicity photo of the IBM SSEC, 1948. Bill McClelland at the table-lookup unit on the left, Betsy Stewart at the console, an engineer on the right. It was famously retouched to not show the prominent black columns supporting the building.

That idea of legibility goes back to the ENIAC, the story of a gigantic computer I bring up in *New Dark Age*. The ENIAC is the moment at which we see computational visibility start to disappear. The ENIAC was this room-sized supercomputer, but it was covered in all these little flashing lights, and people who worked on it said you could follow the process of a piece of computation across the walls in these little lights. And then you think about these examples in the exhibition, the MareNostrum supercomputer, as beautiful as it looks, inside this church, it's completely inscrutable behind glass. And that followed the example of the IBM supercomputer that they built on Fifth Avenue, in New York, in the 1960s. They put this huge supercomputer behind glass, and people would come up and press their noses against it and wonder at the future of technology, not knowing all the time this thing was running calculations for a hydrogen bomb explosion. That is what's happening all the time when we interact with systems that we don't have this visibility into, that we have no understanding of the decisions that are being made with them, on our behalf. Our capacity for action is being reduced by interacting with those systems. It's not just that they may be making decisions on our behalf, but that all the possible things that we might do, that we might do differently, are disappearing without us even being aware of it. That's the disempowering that occurs within this opacity of systems that concerns me, and why I'm so interested in systems that actually promote agency. If you look at something like the new RUSS[4] development in London, along Segal principles of self-building and community ownership, these quite small changes, which are very hard to achieve within the current planning law and legal framework, you see the step change in quality of life. It's kind of extraordinary. And for me, that's entirely about agency, reducing opacity, and giving people the capacity not just to act, but to learn how to act through their interaction with these systems.

Data centers are invisible to the majority of the people that we've talked with. So, we decided that making things visible is a big part of how we curate this exhibition. For us, having agency and being political are two sides of the same coin. Because you can easily become entangled in conspiracy theories if you feel you don't have any political agency, or you don't have any kind of forum to express your opinion. And I think the smart city as we know it today, or this endless datafication where the systems fall in the background, and you just need to trust the flow of this thing, is very dangerous, actually. Because it strips you of any kind of ability to also be political within the creation of these systems—because they are done behind closed doors.

I was thinking of the walk that I did that I described in *New Dark Age*, where I traversed London, walking between two data centers; the London Stock Exchange that's in Slough out to the west of London, and walked all the way through the center of London, following this microwave data link, all the way to the New York Stock Exchange, Essex, out in East London. Seeing where things are, seeing what their context looks like, might be the first lesson in this.

On that particular walk, the two signal moments for me were following the line of structures that support this microwave link as it popped over this NHS hospital near Heathrow, where you had this huge 1970s concrete block NHS hospital, which had been a wonder of healthcare when it was built, and is now, like the rest of Britain's healthcare system, incredibly underfunded, literally crumbling, both in services and architecture. On the top were these two huge microwave dishes through which billions of dollars in transactions were constantly pouring, perched parasitically atop this public social architecture, and operating completely invisibly. No one around that had any idea what these dishes were for, what

Microwave dishes mounted on Hillingdon Hospital, December 2014.

this information was. Still, it was absolutely raw extraction of capital from a broader system that's exemplified by the opacity with which it operated. And that was reinforced, of course, when I got near the New York Stock Exchange data center out in Essex, where, as soon as I appeared, I was immediately approached by security guards and threatened by the police. Because the point of this is opacity, it's designed to be opaque because they are hiding things. Because if you fundamentally understand that there is enough money pouring over the top of that NHS hospital every second of every day to renovate the whole thing and give everyone brilliant patient care, people are going to care about that. This political sense was given such an incredible form by understanding it architecturally and by approaching it physically. I think that the interaction between those two experiences must be really emphasized. Architecture should be a bodily experience. And when it ceases to be bodily and physical, something is wrong.

> In the history of computing, "anachronistic failures," as you put it, are very common. They tend to be written off, feeding a cycle of obsolescence and constant revision. However, you argue that there have been some very interesting innovations along the way. Machines such as Turing's theoretical "oracle machine" or the so-called MONIAC water computer[5] were used to model the British economy. These are types of machines that offered different ways of modeling. The MONIAC offered something manual and low tech: you just needed to add water to the system, and it was able to calculate all these different aspects of the economy.

They're brilliant stories in history. It remains astonishing to me that 99.999 percent of all computers on the planet are one type of computer. When infinite forms of computation are possible, essentially. The example of the MONIAC water computer is interesting for a couple of reasons. One of which is this question of transparency and opacity that we keep coming back to. This was a very accessible machine. It's very weird, the size of a large refrigerator. It's covered in little taps and knobs and pulleys, and it's got all these labels on them, like income tax and VAT and Treasury and imports and exports. By twiddling all these knobs, you can model the British economy. It makes water flow through

Professor A. W. H. (Bill) Phillips with the Phillips Hydraulic Computer or Monetary National Income Analogue Computer (MONIAC), ca. 1958–67.

different pipes and into different pots, so that you really see the granular impact of every decision that you make as it happens right in front of you. And the level of control that this enabled meant that this system, which was originally designed for teaching, ended up being used within the British Treasury. They found it gave economic planners so much more understanding of the system that they were interacting with, which then enabled them to work and use that system better. And that, for me, is absolutely about opacity. It's actually visualizing on a much deeper level what was occurring. I think that's important, because one of the things that the MONIAC acknowledges is that the world itself is not a binary system. The world itself is analogue. It flows, it changes, it's fluid. And modeling in fluids seemed, for a time at least, compared to the kind of computation that was available otherwise digitally at the time, far more powerful to do it in a material that was of the world.

> The modernist mantra has been this specialization and compartmentalization of knowledge. Environmental thinking goes against that, it seems, as we need to think through the environment in ways that maybe we don't know how to do yet.
>
> As spatial planners, we are so focused on creating models that are all about reducing complexity to gain control. Especially in architecture, it seems to come from a modernist way of working, practicing, and thinking about architecture. How can we move a bit away from it? Is there a way to shift away from a reductionist way of creating models about the future of our spaces and ways we inhabit them? Do we need models at all?

Epilogue

The first thing is that the notion of the environment has become an incredibly destructive one. Because it already starts to enforce this distinction, this separation. This idea of figure and ground of the human that is separate from everything that's around, rather than being utterly enmeshed within it. The two are totally separate. So perhaps I would plead for an idea of ecological thinking rather than environmental thinking, because ecology is essentially the science, but also the worldview of interconnection that stresses that it's the relationships that matter, the connections between things, rather than the nodes that actually create meaning, and that, ultimately, nourish everything.

And so any kind of passage toward more ecological ways of thinking has to work against these fixed binaries that so much of the thinking processes are stuck in. It's just a really good example of where you start with a model, and then you try and fit that model into the existing world, rather than understanding what can emerge from a deeper, emergent relationship with the world. Ecological thinking, for me, is a kind of thinking in which many kinds of thinking happen, which is the example of taking into account the ways of other beings, when we make buildings and structures and planning.

> In her book, *Friction* (2004), Anna Lowenhaupt Tsing looks at the machinery of the world. Reading her work made us understand how reductive a lot of these systems are, which we were not aware of before. Moving away from that thinking also seems sometimes obvious, but also very difficult to do because we're so enmeshed in this system. Coming to essential questions and thinking through them over time, and ecologically, is something we also learn from your work.

The term that I increasingly use for myself is "flow," which is actually how I put myself and my work into a much larger current that is already in play, a much larger ecology. It really came home to me in a series of catastrophic events over the last few years, being present on the island I live on. These extreme weather events cause huge damage. In *Ways of Being*, I wrote a bit, and perhaps quite naively, about things like "the Half Earth Hypothesis," which largely argues that humans need to abandon large areas of the Earth in order for natural processes to be allowed to take their place again. But that just doesn't really apply here. The wildfires that have raged around Athens, for example, in the last few years, are largely driven by our abandonment of the land, which has become depopulated. The land on which people have lived for thousands and thousands of years is no longer being tended in ways that previously would have suppressed the wildfires.

Of course, they're still being driven by rising temperatures, but they're also a result of changes in human activity. Likewise, where I live, when you get these large storms, you see the soil being washed into the sea, because the old systems of water maintenance and water management from the large water tanks that used to be dug to retain seasonal water through to the terraces, which used to protect farmland and are mostly now falling down, those aren't maintained anymore. So, what does it mean to come back into the flow of these ecological systems? Well, it means to tend to the land in various ways.

Now, I'm talking about myself in a quite rural contexts, but what does it mean to come into the flow of larger ecological systems within cities or within housing? It's understanding that one is not the beginning of any of these processes. One has a very small part within a far larger chain of processes that are occurring. We need to acknowledge that we are not the river, but that we can rearrange a few of the stones in it. So, what can we do by moving those stones around? That is reflective of and supportive of a greater flow that is occurring. And that's how I mostly try to think of what I'm trying to do now.

1. Liping Zhu, Song-Ju Kim, Masahiko Hara, and Masashi Aono, "Remarkable Problem-Solving Ability of Unicellular Amoeboid Organism and Its Mechanism," *Royal Society Open Science,* December 19, 2018, https://royalsocietypublishing.org/doi/10.1098/rsos.180396.

2. "How Clams Help Keep Polish Water Clean," *The Economist,* January 21, 2021, https://www.economist.com/europe/2021/01/21/how-clams-help-keep-polish-water-clean.

3. Alice Grahame, "'This Isn't at All Like London': Life in Walter Segal's Self-Build 'Anarchist' Estate," *The Guardian,* September 16, 2015, https://www.theguardian.com/cities/2015/sep/16/anarchism-community-walter-segal-self-build-south-london-estate.

4. Oliver Wainright, "'I Plumbed in Our Bath – and It Works!' The DIY Diehards Who Built 36 Affordable Homes from Scratch," *The Guardian,* June 4, 2024, https://www.theguardian.com/artanddesign/article/2024/jun/04/plumbed-bath-diy-diehards-affordable-homes.

5. The Monetary National Income Analogue Computer (MONIAC) was built in 1949 by economist Bill Phillips while he was a student at the London School of Economics. It was able to model the entire economy decades before digital computers became common. See James Bridle, *Ways of Being* (London: Penguin Books, 2023).

Appendix

Biographies

James Bridle is a writer, artist, and technologist. Bridle holds a master's degree in computer science and cognitive science from University College London, where their dissertation focused on the creative applications of artificial intelligence.

Bridle's artworks have been commissioned by galleries and cultural institutions and exhibited internationally as well as online. Notable works include *Seamless Transitions* (The Photographers' Gallery, London, 2015), *The Glomar Response* (NOME Gallery, Berlin, 2015), and *My Delight on a Shining Night* (Filodrammatica Gallery, Rijeka, 2018). Their writing has appeared in magazines and newspapers including *Wired*, *The Atlantic*, *New Statesman*, *The Guardian*, and the *Financial Times*. They are the author of *New Dark Age* (2018) and *Ways of Being* (2022), and they wrote and presented *New Ways of Seeing* for BBC Radio 4 in 2019.

Giulia Bruno is a Berlin-based artist working with photography, video, language, and media. Holding a master's degree in biology from the Università degli Studi di Milano and master's degrees in both photography and filmmaking, she explores the intersections of technology, nature, cultural activism, and language, with a focus on climate change, biodiversity, and artificial landscapes.

Bruno's work has been exhibited in various international shows, including *Earth Indices* at Haus der Kulturen der Welt in Berlin (2022), *Matar la Nube* at the Plomo Gallery in Mexico City (2022), and *Uncanny Values: Artificial Intelligence & You* at the MAK – Austrian Museum of Applied Arts in Vienna (2019). Her documentary *Capital* received the Visioni Italiane award in 2015. She co-founded Studio Poor with Paola Raheli and currently collaborates with artist and musician Giuseppe Ielasi on an audiovisual performance for the new exhibition space Voce at Triennale Milano and upcoming tours, working on the concepts of materiality in image and sound.

Teresa Fankhänel is a curator, writer, and editor in chief of the *Architectural Exhibition Review* based in Karlsruhe. She holds a master's degree from the Bartlett School of Architecture and a PhD from the University of Zurich, and she is currently acting professor of architecture theory at the Karlsruhe Institute of Technology.

Her recent exhibitions include *Shouldn't You Be Working?* (2023), *Andrea Canepa: As We Dwell in the Fold* (2023–24), and *Farmland* (2025). Her interests include the use of technologies and media in architectural design, as well as the history, theory, and practice of architecture exhibitions. At the Architekturmuseum der TUM, her large-scale research project Pixels, Vectors and Algorithms, which explored the origins of digital tools in architecture, resulted in the exhibition and accompanying catalogue *The Architecture Machine: The Role of Computers in Architecture* (2020).

Cara Hähl-Pfeifer is a curator at the Architekturmuseum der TUM and a research associate at the TUM Chair of Architecture History and Curatorial Practice. She holds a master's degree in architecture from the Technical University of Munich and previously studied and worked in Mendrisio, Berlin, and Karlsruhe.

She has held various positions in architecture, research, and curatorial practice, combining her interests across multiple disciplines and media. In 2023, she joined the Architekturmuseum der TUM as a curatorial assistant for the exhibition project *The Gift: Stories of Generosity and Violence in Architecture*. Independently, her work investigates socially and ecologically inequitable urban environments in a post-migrant society through anthropological and historical analysis.

Max Hallinan is a software engineer based in New York City. He currently works as a generalist engineer at Mercury, a banking platform for startups, and holds a bachelor's of fine arts from the Maryland Institute College of Art. Hallinan is interested in the cultural and professional implications of machine-assisted programming, having started his career in the era of handwritten software.

He is currently collaborating with Teresa Fankhänel and the Architekturmuseum der TUM on a research project with the Karlsruhe Institute of Technology, investigating the role of text-based AI in architectural archives and exhibition design.

Mél Hogan is the host of *The Data Fix* podcast and editor of *Heliotrope*, a space for scholars and practitioners to explore and share their work on the intersection of culture, media, technology, and the environment. Hogan is based in Kingston, Canada, where she is an associate professor in the Department of Film and Media at Queen's University. She holds a PhD from Concordia University.

Her research focuses on environmental media and data infrastructure in the contexts of planetary catastrophes and collective anxieties about the future. Hogan has published in *New Media & Society*, *Work Organisation, Labour & Globalisation*, *Ephemera*, and other journals.

Catherine Hyland is an artist living and working in London. She holds a bachelor's degree from Chelsea College of Art and Design and a master's degree from the Royal College of Art. Her photography centers around people and their connection to the land they inhabit.

Primarily landscape based, Hyland's work is rooted in notions of fabricated memory, grids, enclosures, and national identity and has been exhibited at Month of Photography Los Angeles, Renaissance Photography Prize, National Portrait Gallery, and Design Museum in London, among other places. Her ongoing projects highlight humanity's attempts—some more effective than others—to tame and transform nature, both past and present.

Damjan Kokalevski is a curator at the Architekturmuseum der TUM and postdoctoral researcher at the TUM Chair of Architecture History and Curatorial Practice. He earned his PhD in 2018 from ETH Zurich with a dissertation titled *Performing the Archive: Skopje – From the Ruins of the City of the Future*. His doctoral thesis was further developed into a series of curatorial projects and the book *Skopje Walkie Talkie*, co-edited with Susanne Hefti (2019).

Born and raised in Skopje, he trained as an architect and has worked across academia and practice in Vienna, Tokyo, Zurich, and Munich. His work centers on curatorial inquiries into archival knowledge, viewed through the lens of social justice, democratization of knowledge, accessibility, and resources. Most recently, Kokalevski co-curated the exhibition *The Gift: Stories of Generosity and Violence in Architecture* at the Architekturmuseum der TUM (2024) and co-edited *the e-flux architecture* online publication *The Gift*.

Andres Lepik is the director of the Architekturmuseum der TUM and Professor of Architecture History and Curatorial Practice at TUM. He holds a graduate degree in art history and a PhD, with a dissertation on Renaissance architectural models.

After working as a curator at the Neue Nationalgalerie in Berlin and in the Architecture and Design Department of the Museum of Modern Art in New York—where he was responsible for the exhibition *Small Scale, Big Change: New Architectures of Social Engagement* (2010)—he became a Loeb Fellow at the Graduate School of Design at Harvard University. Among his exhibitions at the Architekturmuseum der TUM in Munich are *AFRITECTURE: Building Social Change* (2013–14), *Who's Next? Homelessness, Architecture, and Cities* (2022–23), and, most recently, *Visual Investigations: Between Advocacy, Journalism, and Law* (2024–25).

Niklas Maak is a writer, educator, and editor for architecture at the *Frankfurter Allgemeine Zeitung*, based in Berlin. He studied art history, philosophy, and architecture in Hamburg and Paris and holds a PhD in art history from Hamburg University. Maak has taught architectural history and theory at Harvard University and served as a visiting professor at the Städelschule in Frankfurt.

His publications include *Le Corbusier: The Architect on the Beach* (2010) and *Technophoria* (2020). Maak's *Server Manifesto* (2022) focuses on questions of data ownership, digital sovereignty, and urbanism in the digital age. In 2022, he was granted an artist program stipend of Villa Massimo in Rome. Maak has received numerous awards for his writing, including the George F. Kennan Commentary Prize, the BDA Prize for Architecture Criticism, the Henri-Nannen Prize, and the Johann-Heinrich-Merck Prize for literary criticism and essay.

Marija Marić is an architect and researcher based in Luxembourg. She works as a postdoctoral research associate at the University of Luxembourg, where she also teaches. In 2020, she earned her doctoral degree from the Institute for the History and Theory of Architecture at ETH Zurich.

Her dissertation, titled *Real Estate Fiction*, critically examined the role of communication industries in the design, mediation, and commodification of housing. Marić co-curated the Luxembourg Pavilion at the 2023 Venice Architecture Biennale and is the co-author of *Staging the Moon: Resource Extraction Beyond Earth* (2023). Her work—centered on questions of property, housing, and land, and viewed through the lenses of care and spatial justice—has been exhibited, presented, and published internationally.

Anna-Maria Meister currently directs the Lise Meitner Research Group "Coded Objects" at the Kunsthistorisches Institut in Florenz – Max Planck Institut, and is a full professor of Architecture Theory and co-director of the saai archive at the Karlsruhe Institute of Technology. She is a licensed architect with a PhD from Princeton University as well as architecture degrees from Columbia University and the Technical University of Munich.

Meister's work across medial divides, including exhibitions and installations, focuses on processes of design and the design of processes, the materiality of knowledge systems, history of ideas and methodological experimentation. Her recent publications include the co-edited *Radical Pedagogies* (2022), *Entangled Temporalities* (2023), and *Are You a Model?* (2024); she is currently completing a monograph titled *Nation of Norms: Designing German Worldviews One Object at a Time*.

Marina Otero Verzier is Dean's Visiting Assistant Professor at GSAPP, Columbia University, in New York City, where she leads the Data Mourning clinic, exploring the intersection of digital infrastructures and climate catastrophe. She holds a PhD in architecture, earned in 2016 from ETSAM (Escuela Técnica Superior de Arquitectura de Madrid).

A 2022 Harvard Wheelwright Prize winner, she has collaborated with institutions such as the DIPC Supercomputing Chile's Ministry of Science on projects addressing data infrastructures and extractivism. Otero authored *En las Profundidades de la Nube* (2024), and previously led the MA Social Design program at Design Academy Eindhoven (2020–23) and directed research at Het Nieuwe Instituut (2015–22). Her curatorial work includes *Wet Dreams* (2024), *Compulsive Desires* (2023), and *Work, Body, Leisure*, the Dutch Pavilion at the 2018 Venice Architecture Biennale. Otero has co-edited several publications, including *Automated Landscapes* (2023), *Lithium: States of Exhaustion* (2021), and *More-Than-Human* (2020).

Trevor Paglen is a New York City–based artist whose work spans image-making, sculpture, investigative journalism, writing, engineering, and numerous other disciplines. He earned a master of fine arts from the School of the Art Institute of Chicago and later completed a PhD in geography at the University of California, Berkeley.

Paglen has launched an artwork into distant orbit around Earth in collaboration with Creative Time and MIT in 2012 (*The Last Pictures*) and created a radioactive public sculpture for the exclusion zone in Fukushima, Japan, in 2015 (*Trinity Cube*). He has had solo exhibitions at the Smithsonian Museum of American Art in 2018–19 (*Sites Unseen*), the Fondazione Prada in 2019–20 (*Training Humans*), and the Barbican Centre, London, in 2019–20 (*From "Apple" to "Anomaly"*), among others. He also participated in group exhibitions at the Metropolitan Museum of Art, the San Francisco Museum of Modern Art, the Tate Modern, and numerous other venues.

Godofredo Enes Pereira is a London-based architect and researcher who serves as Head of Program for the MA Environmental Architecture at the Royal College of Art. He holds a master's degree from the Bartlett School of Architecture and a PhD from the Centre for Research Architecture, Goldsmiths University, and is focused on publishing and exhibiting on environmental architecture and collective politics.

His doctoral thesis investigated political and territorial conflicts within the planetary race for underground resources; at Forensic Architecture, he led an investigation of environmental and human rights violations in the Atacama Desert. He is currently working on the publication of *Ex-Humus*, developing research on the Lithium Triangle across Chile, Bolivia, and Argentina, and is a co-investigator on the project Scales of Climate Justice, funded by the British Academy.

Andra Pop-Jurj is an advanced researcher at Forensic Architecture in London, where she investigates state and corporate violence through spatial analysis and event reconstruction, with a focus on environmental toxicity and injustice, often perpetrated by the petrochemical industry. Pop-Jurj holds a master's of arts in architecture from the Royal College of Art.

Her independent practice explores the material and political dimensions of extractivism and environmental degradation in high-latitude regions. She researches across disciplines and media, integrating writing with digital techniques—a practice she frequently shares through lectures and workshops. Situated at the intersection of digital arts and science, her most recent project, Monsters and Ghosts of the Far North, was developed collectively and has been exhibited internationally as a multiplayer cartographic installation.

Alison Powell is a writer and associate professor in media and communications at the London School of Economics, where Powell directs the MS stream in Data & Society. Powell holds a PhD in communication studies from Concordia University and researches knowledge politics and technology innovation, with a focus on participatory and inclusive design.

Recent projects include the JUST AI network, which diversified data and AI research in the United Kingdom, and the book *Undoing Optimization: Civic Action and Smart Cities* (2021). Powell's practice-based research includes Data Walking—a participatory method for exploring data in urban environments—and an interactive tool and workshop format called *Prototypes for Ethical Reflection*. A current writing project, *Deceptive Stories of Technology (And What to Do About Them),* explores how technology and knowledge might be valued differently.

Māra Starka is part of the curatorial team for the *City in the Cloud – Data on the Ground* exhibition project at the Architekturmuseum der TUM. She holds a master's of arts in Architecture from TUM and works as a practicing architect in Riga, Latvia.

Originally from the Baltics, her work is shaped by a sustained interest in infrastructural connectivity and the ecological and social imaginaries of digitalization. Her experience spans recognized public space renovation projects, educational initiatives, and scenography. In addition to her curatorial practice, she has contributed to teaching in the architecture program at TU Munich.

Rafael Brundo Uriarte received his PhD in computer science in 2015 from the IMT School for Advanced Studies in Lucca, Italy. He coordinates the Digital Humanities Laboratory (DH Lab) at the Kunsthistorisches Institut in Florence, where he currently manages over fifteen DH projects related to art history.

He has won several prestigious grants like the TU Wien's Marie Skłodowska-Curie Individual Fellowship and the Fondazione CON IL SUD's Brains to the South at the University of Cagliari. Alongside his work on the evolution of AI, machine learning, blockchain, and cloud computing, he has been actively engaged in digital humanities for over ten years. Notable projects include a digital platform for Venetian Music and the digitization of seventeenth-century choir books from St. Mark's Basilica, Venice.

Image Credits

Cover (top): Catherine Hyland, 2018
Cover (bottom): Giulia Bruno / Architekturmuseum der TUM, 2025
19: Virgina Zangs, 2025
20: Marina Otero Verzier, 2025
21: Marina Otero Verzier, 2023
22 (top): freedom_wanted / Alamy Stock Photo, 2023
22 (bottom): Catastro de Concesiones Mineras, https://catastro.sernageomin.cl
25 (top and bottom): Deutsches Museum
27 (top): Bildarchiv der Deutschen Kolonialgesellschaft, Universitätsbibliothek Frankfurt am Main
27 (bottom)–28: Rudolf Schlechter, *Die Guttapercha- und Kautschuk-Expedition des Kolonial-Wirtschaftlichen Komitees, wirtschaftlicher Ausschuss der Deutschen Kolonialgesellschaft nach Kaiser Wilhelmsland 1907–1909* (Berlin 1911)
29: Wikimedia Commons: https://de.m.wikipedia.org/wiki/Datei:Carte_générale_des_grandes_communications_télégraphiques_du_monde.jpg
29: National Maritime Museum, Greenwich, London
31 (top): Deutsches Museum
31 (bottom): Siemens Historical Institute
32: Siemens Historical Institute
33: Antiqua Print Gallery / Alarmy Stock Photo
34: Library of Congress, https://www.loc.gov/item/93510355/
34–37: Deutsches Museum
38: Library of Congress, https://www.loc.gov/item/99614178/
39–41: Trevor Paglen. Courtesy of the Artist, Altman Siegel, San Francisco and Pace Gallery
51–67: Catherine Hyland, 2018
69–70: Godofredo Enes Pereira
71 (top): Grupo de Investigação Territorial (GIT), 2022
71 (bottom)–72: Godofredo Enes Pereira
77: Cara Hähl-Pfeifer / Architekturmuseum der TUM, 2025
78: Maria Heinrich, 2022. Commissioned by Het Nieuwe Instituut and first published in *Automated Landscapes* (2023), edited by Marina Otero Verzier, Merve Bedir, Ludo Groen, Marten Kuijpers, and Víctor Muñoz Sanz
79: Barcelona Supercomputing Center
80 (top): Leibniz Supercomputing Centre (LRZ)
80 (bottom): Helena Francis, 2023. Created for the Future Storage project by Marina Otero Verzier as part of her Wheelwright Prize research
89: Courtesy of Steve Jones Flight by Southwings for the Southern Environmental Law Center
90 (top): Twitter. First published in Niklas Maak, *Server Manifesto: Data Center Architecture and the Future of Democracy* (Berlin: Hatje Cantz, 2022)
90 (bottom): US Army Photo. First published in Niklas Maak, *Server Manifesto: Data Center Architecture and the Future of Democracy* (Berlin: Hatje Cantz, 2022)
91 (top): First published in Niklas Maak, *Server Manifesto: Data Center Architecture and the Future of Democracy* (Berlin: Hatje Cantz, 2022)
91 (bottom): CYBERSYN/Cybernetic Synergy. First published in Niklas Maak, *Server Manifesto: Data Center Architecture and the Future of Democracy* (Berlin: Hatje Cantz, 2022)
92 (bottom): Courtesy of Snøhetta/Plompmozes. First published in Niklas Maak, *Server Manifesto: Data Center Architecture and the Future of Democracy* (Berlin: Hatje Cantz, 2022)
93: schneider+schumacher. First published in Niklas Maak, *Server Manifesto: Data Center Architecture and the Future of Democracy* (Berlin: Hatje Cantz, 2022)
94: Niklas Maak and Stefan Sauter. First published in Niklas Maak, *Server Manifesto: Data Center Architecture and the Future of Democracy* (Berlin: Hatje Cantz, 2022)
97–113: Giulia Bruno / Architekturmuseum der TUM, 2025
115: Alison Powell
117 (top): Richard Galpin, Alberta Fruit Commons
117 (bottom): R. Griffiths, "Sensing the Luminous Night: Capturing and Communicating Time-based Observations of Environmental Light Across Urban and Rural Sites." In J. Han, D. Lombardi, and A. Cece, eds., *Advances in the Integration of Technology and the Built Environment*, AAB 2024, Lecture Notes in Civil Engineering, vol 593 (Singapore: Springer, 2005), pp. 90–97
118: Alison Powell
123: Tabakalera, 2024
124: Herzog & de Meuron, 2017
125: Viksha Nayak and Vishesh Sahni / Columbia GSAPP, 2025
126: Helena Francis, 2021. Developed within the ADS8 studio Data Matter: Digital Networks, Data Centres, and Post-Human Institutions at the Royal College of Art, led by Ippolito Pestellini, Marina Otero Verzier, and Kamil Dalkir
131–38: Andra Pop-Jurj and Lena Geerts Danau
143: Nicholas Negroponte, The Architecture Machine: Toward a More Human Environment (Cambridge, MA: MIT Press, 1970), figure 10, p. 85
144: Courtesy of the Jencks Foundation at the Cosmic House
145 (left): Alexander Mordvintsev, 2015
145 (right): Joern Ploennigs and Markus Berger / Universiy of Rostock
159–171: Giulia Bruno / Architekturmuseum der TUM, 2025
172–189: Giulia Bruno
193: A. Tero et al., "Rules for Biologically Inspired Adaptive Network Design," *Science* (2010), https://www.science.org/doi/full/10.1126/science.1177894
195: Architectural Press Archive / RIBA Collections
196: Columbia University, https://www.columbia.edu/cu/computinghistory/ssec.html
197 (left): James Bridle, 2014
197 (right): LSE Library, no known copyright restrictions

The book is published on the occasion of the exhibition *City in the Cloud – Data on the Ground* at the Architekturmuseum der TUM in the Pinakothek der Moderne, Munich, October 16, 2025–March 8, 2026.

Director: Andres Lepik
Curator: Damjan Kokalevski
Research Advisor: Marina Otero Verzier
Research and Photography: Giulia Bruno
Photography: Catherine Hyland
Curatorial Assistants: Ramona Kornberger, Leo Paulmichl, and Māra Starka
Public Program Coordinator: Sarolta Szatmári

Exhibition Design: CPWH, Munich
Graphic Design: WVH – Wiegand von Hartmann, Munich
Interactive Design: 3e8 Studio, Vienna

Copyediting of the Exhibition Texts: Anna J. Barańczak (English) and Julika Zimmermann (German)
Technical Exhibition Planning and Execution: Andreas Bohmann, Volker Enders, and Thomas Lohmaier
Conservator: Anton Heine
Registrar: Thilo Schuster
Administration: Martina Heinemann, Rosa Anna Perrini, and Rike Menacher
Public Relations: Martina Heinemann and Lisa Clausen-Schaumann
Press: Cara Hähl-Pfeifer, Julia Kaufmann and Tine Nehler
Freundeskreis Architekturmuseum TUM: Dietlind Bachmeier
Exhibition Prints: ESCHER Digitaldruck, Gersthofen

The exhibition is supported by PIN. Freunde der Pinakothek der Moderne e.V. and their cooperation partner Allianz.
Cooperation Partners: Deutsches Museum; Digital Twin Munich by the City of Munich, Department of Communal Services, and GeodataService; Leibniz Supercomputing Centre of the Bavarian Academy of Science and Humanities
Corporate Sponsor: beMatrix Germany GmbH
Additional Funding to Support the Publication: Nemetschek Group, Schnitzer& GmbH

Acknowledgments

The curatorial and editorial team would like to express their sincere thanks to all collaborators and contributors whose work led to the successful completion of the book and exhibition.

Special thanks go to the architects, artists, and researchers who contributed to the exhibition content: Kate Crawford, Lena Geerts Danau, Kathrin Dörfler, Teresa Fankhänel, Helena Francis, Max Hallinan, Maria Heinrich, Christoph Ignaz Kirmaier & 3e8, Ilina Kokaleska, Ema Krakovska, Romea Muryń & Locument, Godofredo Enes Pereira, Andra Pop-Jurj, Mario Santamaría, Begum Saral, and Virginia Zangs.

Additional thanks go to our exhibition partners and supporters: Luise Allendorf-Hoefer, Mathieu Bujnowskyj, Niklas Dienst, Lance Firmhofer, Moritz Heber, Daniel Kolb, Korbinian Kringer, Elisabeth Mayer, Markus Mohl, Rainer Oesmann, Saakib Sait, Martin Schnitzer, Sabrina Schulte, and Klaus Stegmaier.

We would also like to thank the students of the Technical University of Munich who took part in our seminars and contributed with their research and exhibition projects: Yuval Ehud, Bruno Heringer, Raqida Kasworo, Ulrich Kneisl, Yu Yuan Ko, Weronika Kłósek-Gniewkowska, Paula Löffler, Ramona Lutz, Nora Maleh, Geraldus Martimbang, Lukas Meisner, Günter Merk, Ekaterina Pestriakova, Rosa Schuster, Ghania Syed, Tuan Pham, Viktoria Yurzinova, and Anika Zeman.

Scan the QR code in the book to access additional material and exhibition views. The content behind these BOOK+ QR codes will grow over time. Rather than being stored on web servers or in the cloud, it is stored on a blockchain database to prevent data attacks and manipulation while guaranteeing access security.

Colophon

Editors: Cara Hähl-Pfeifer, Damjan Kokalevski, and Andres Lepik
Project Management: Lisa Luksch and Cristina Steingräber
Copyediting: Aaron L. Bogart
Translation into English (text by Niklas Maak): Anna Bröderdörp
Interview Transcription: Yuval Ehud
Graphic Design: WVH – Wiegand von Hartmann, Germany
Production: Sonja Bröderdörp
Reproductions: Optische Werke Hamburg GbR, Germany
Printing and Binding: Memminger MedienCentrum Druckerei und Verlags-AG, Germany

© 2025 Architekturmuseum der Technischen Universität München (TUM),
ArchiTangle GmbH, and the contributors

Architekturmuseum der TUM
in der Pinakothek der Moderne
Barer Str. 40
80333 Munich
Germany
www.architekturmuseum.de

ArchiTangle GmbH
Meierottostraße 1
10719 Berlin
Germany
www.architangle.com

ISBN 978-3-96680-038-9

All rights reserved; the contents of this publication may not be reproduced, stored in a retrieval system, or transmitted without the prior written permission of the editors and the publisher.

TUM

A.M.